FIELD TRIP GUIDE BOOK 6
1983

A guide to the GLACIAL LANDSCAPES OF DANE COUNTY, WISCONSIN

by

David M. Mickelson

Department of Geology and Geophysics
University of Wisconsin-Madison

University of Wisconsin-Extension
GEOLOGICAL AND NATURAL HISTORY SURVEY
Meredith E. Ostrom, Director and State Geologist

Published by and available from:
University of Wisconsin-Extension, Geological and Natural History Survey
1815 University Avenue, Madison, Wisconsin 53705-4096

To Jean, Amy, Rebecca, and John
for their patience with a field geologist.

ISBN 0-88169-000-7

Cover: False color infrared Landsat Satellite image of Dane County, Wisconsin, May 9, 1976. Courtesy of U. S. Geological Survey, Wisconsin Water Resources District.

TABLE OF CONTENTS

ACKNOWLEDGEMENTS

Many people assisted with the manuscript. Carol McCartney made very helpful comments on the manuscript and made a major contribution to the interpretation of the glacial geology of the County. I would also like to thank reviewers of the manuscript, and especially those that checked the field trip, including Lee Clayton, Nancy Dott, Rosemary Fleming, Francis Hole, Marilyn Hurst, Galen Kenoyer, James Knox, George Knudsen, Louis Maher, Wesley Martin, M. E. Ostrom, Maryanne Stephenson and Richard Whittecar.

Numerous others including John Dallman, John Brown and Ellen Smith provided information for the report. I appreciate their assistance. I also greatly appreciate the fine contribution on soils by Francis Hole. Finally, Mike Czechanski, Judith Lee, Kathie Zwettler, Rose Scott, Michelyn Hass, and Ann Bauhs deserve a great deal of credit for drafting the figures, typing, developing the format for the text and seeing to production of the manuscript. Partial support for field studies was provided by the Wisconsin Alumni Research Foundation and the University-Industry Research Program.

ACKNOWLEDGMENT

[illegible]

A guide to the

Glacial Landscapes of Dane County, Wisconsin

INTRODUCTION

Dane County has a wide variety of landscapes. In the southwest part of the county, East Blue Mound (and the higher West Blue Mound just outside the county) rises 750 feet above the Wisconsin River (Figure 1). In the central part of the county, moraines, bedrock hills and a system of lakes surrounded by flat plains lie in a regional depression. In the northern and eastern parts of the county, numerous drumlins streamlined by the flow of glacier ice produce a northeast-southwest lineation in the landscape.

This book, accompanied by the map, Glacial Geology of Dane County, Wisconsin (in the back pocket), explores the origins of these landscapes. Why are the landscapes different? When did they form? Why are landscapes and materials important to people living in the county?

Two field trip guides designed to explain landscape features are included at the back of this book and the field trip routes are shown inside the front and back covers. By driving or bicycling these routes and following the text and map included in the pocket, readers can enhance their knowledge and enjoyment of Dane County. Technical terms used in this report are defined in the glossary, and I encourage readers not familiar with some terms to use the glossary to help understand the text and map. References to many original papers are given throughout the text so that certain topics can be pursued in more depth.

The oldest glacial deposits present on the surface in Dane County are mapped as unit pgm(pre-Woodfordian ground moraine) on the map, Glacial Geology of Dane County, Wisconsin (in back pocket) and they are present directly east and north of Belleville. These deposits evidently are covered by younger deposits to the north in the vicinity of Verona, but they cannot be traced further to the north. There is no way to date these deposits, but their age is estimated as being early Wisconsin or about 30,000 to 80,000 years old. The materials left by this glaciation are very thin over bedrock and in a few places they lie over even older glacial deposits.

The remainder of the glacial deposits in the county are of Woodfordian or late Wisconsin age. This means that they range from about 14,000 to 20,000 years old and most of them probably are about 15,000 years old. Although the exact time is not well established, it was probably about 15,000 or 16,000 years ago when glacier ice reached its maximum along the glacial border shown by the Johnstown Moraine.

This moraine is a nearly continuous ridge, running from the town of Brooklyn northwestward through Verona and Cross Plains. From Cross Plains to the Wisconsin River the glacial ice margin is marked only by a scatter of glacial debris. The ice that left this debris was part of the Green Bay Lobe which flowed to the southwest from the Green Bay lowland.

The most profound effect on the appearance of landscapes in Dane County was glaciation, which probably occurred several times during the last million years. Earlier geologic events also played an important role in the development of our landscape. For several hundred million years the rock in southern Wisconsin has been exposed to weathering and erosion by streams. During this time huge quantities of

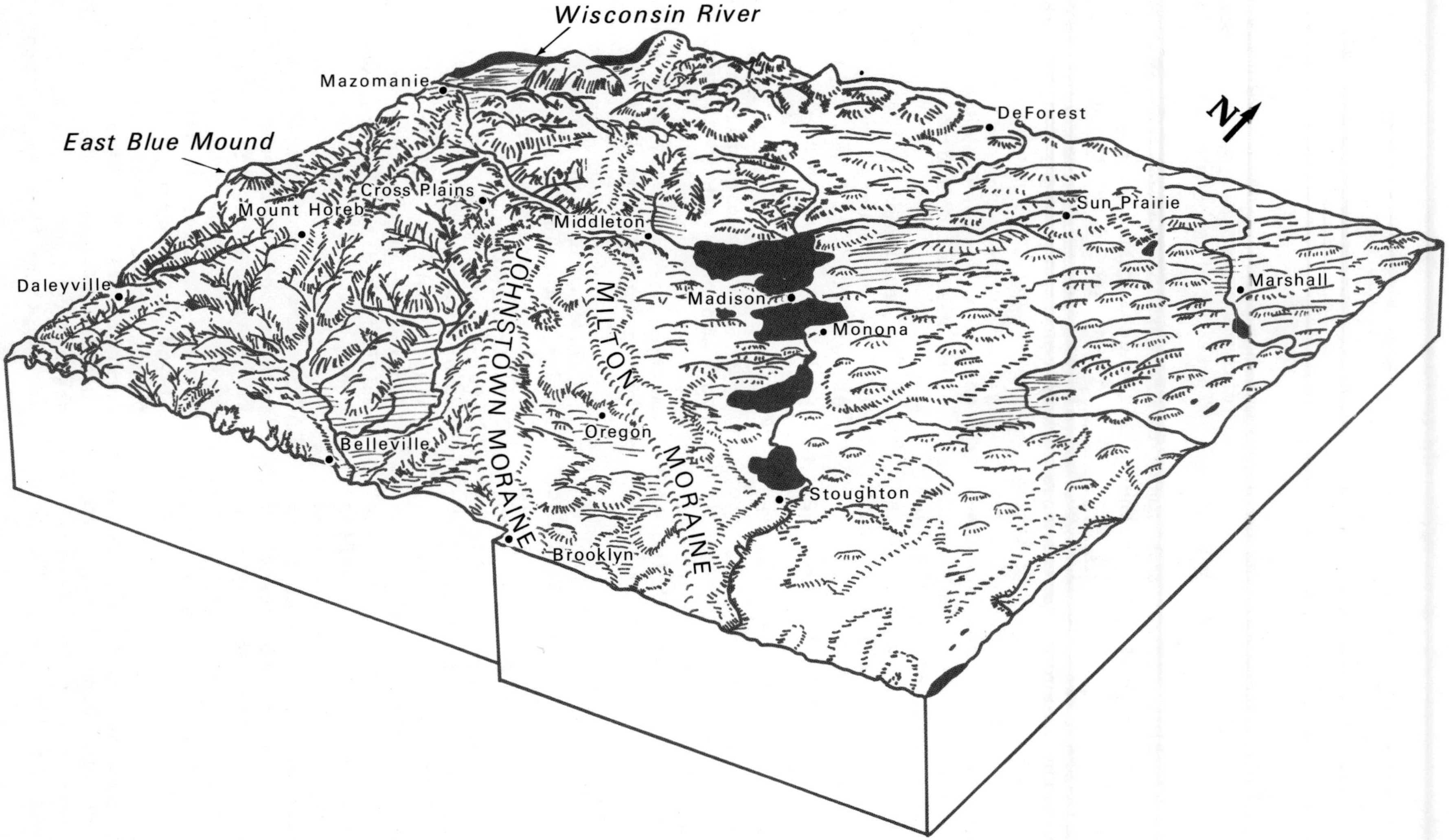

Figure 1: Physiograpahic diagram showing landform distribution in Dane County. Landforms and geology are diagrammatic and individual hills and minor valleys may be missing. (Diagram by Francis Hole and Judith Lee).

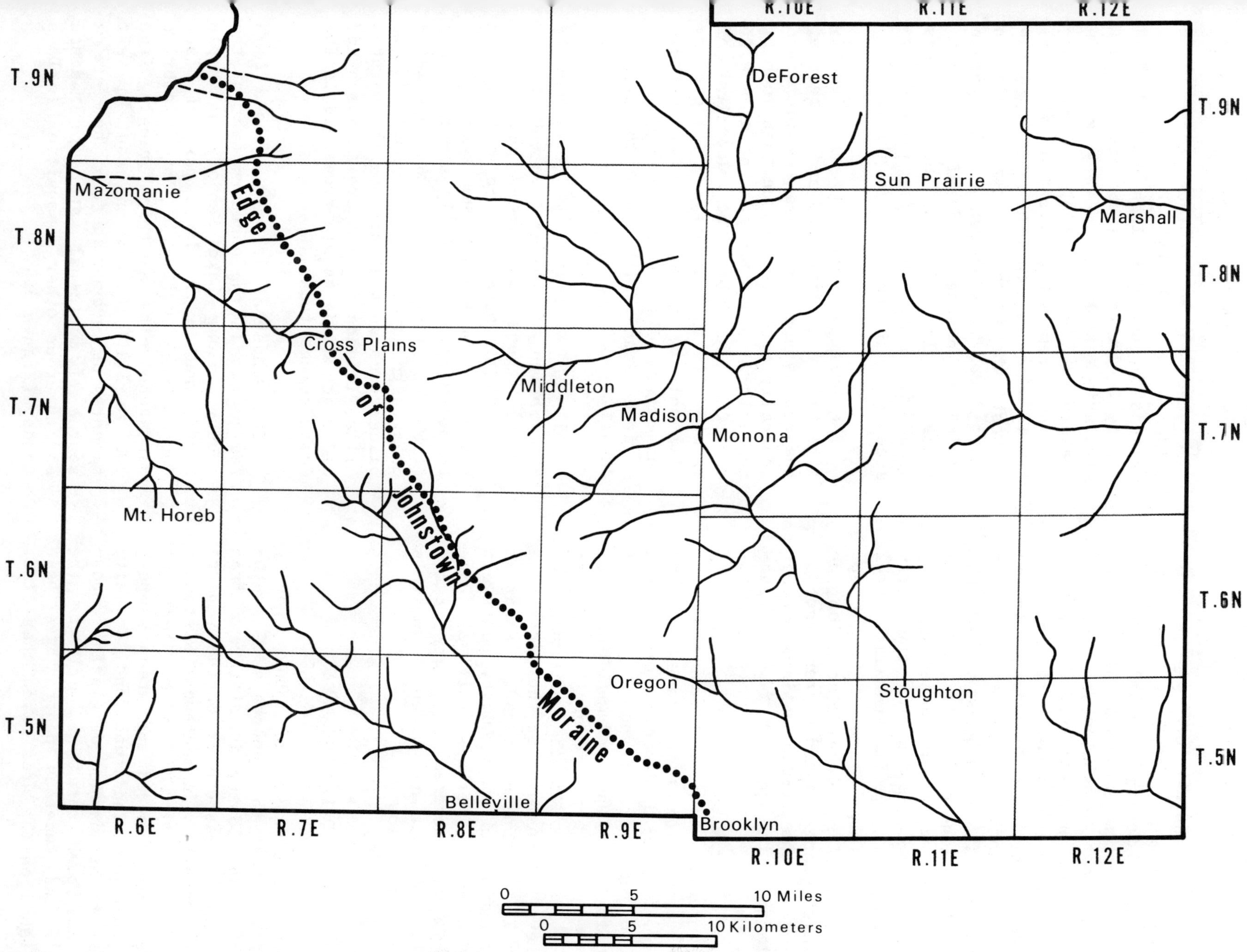

Figure 2: Pattern of major streams which existed in Dane County previous to glaciation. Note that the Yahara River system was extensive and drained much of the central part of the county and that lakes were absent.

rock and soil material were removed from the land surface exposing the sandstone and dolomite that we now see quite commonly in southwest Wisconsin. At the beginning of glacial or Quaternary time, (1 to 3 million years ago), Dane County probably had a drainage system somewhat similar to what we see in southwest Wisconsin today. The bedrock surface (Figure 2), much of which is now buried by glacial deposits, has a deep valley in the Yahara Lowland. Some evidence in northern Illinois suggests that the preglacial landscape did not have valleys as deeply entrenched as they are now. Sometime during the early part of Quaternary time (perhaps 1 or 2 million years ago) time, a slight uplift of the land, local glaciation, or sea level change caused entrenchment or cutting of the stream valleys to approximately the present bedrock surface.

We really have no idea when the first glaciation took place in Dane County. Glacial deposits older than 700,000 years are present in parts of Illinois, Iowa and perhaps Minnesota. It is likely that some of these deposits also exist in the subsurface of southern Wisconsin or at the surface in central Wisconsin (Wood, Marathon Counties) or in Green County southwest of Monroe[2]. Perhaps part of Dane County was covered by very early glacier ice but no evidence for this has ever been found.

THE DRIFTLESS AREA

Most residents of Dane County recognize that the landscape of southwestern Dane County is different from that in the eastern or northern parts of the county. In southwest Wisconsin there are several hundred feet of relief, many wooded and steep slopes, and numerous sandstone and dolomite cliffs. The absence of glacial deposits has been noted as early as the 1820's and numerous papers have been published on the area since that time. Martin[3] presents a detailed history of the discovery and explanation of the Driftless Area. Recently, it was suggested[4] that the Driftless Area was glaciated. To my knowledge there is no evidence of glacial drift on the uplands in the driftless part of Dane County[5]. While it is conceivable that glaciation took place very early in Quaternary time and that all evidence has been removed by erosion, it seems likely that, as in other parts of the midwest where very early glaciation took place, some evidence would remain behind. Because no irrefutable evidence of glaciation in the Driftless area has been found, few people have accepted that the area was glaciated.

The Driftless Area was not directly affected by glacier ice, and therefore, the landscape is less subdued and generally more picturesque than much of the rest of the county. The area also contains few of the deposits of interest in this report and its geology will only be summarized here. Ostrom and others discuss the geology of the sandstones and dolomites of the area in detail.

All of Wisconsin is underlain by Precambrian age (older than 600 million years) igneous and metamorphic rocks. In Cambrian time (500 to 600 million years ago) a shallow sea spread south to north across a gently sloping plain of these older rocks, and sand-later changed to sandstone-was deposited. In places such as what are now the Baraboo Hills, islands stood above this sea and old beach deposits are preserved in the Cambrian age rocks[7].

The sandstone is only exposed in a few places in the northern part of the county, along Lake Mendota at Maple Bluff, and in the lower part of Black Earth Creek valley. These Cambrian sandstone units are thick, ranging from 300 to 1,100 feet[8], and contain vast amounts of groundwater. They are a source of municipal water for Madison and numerous other communities in Dane County (Figure 3).

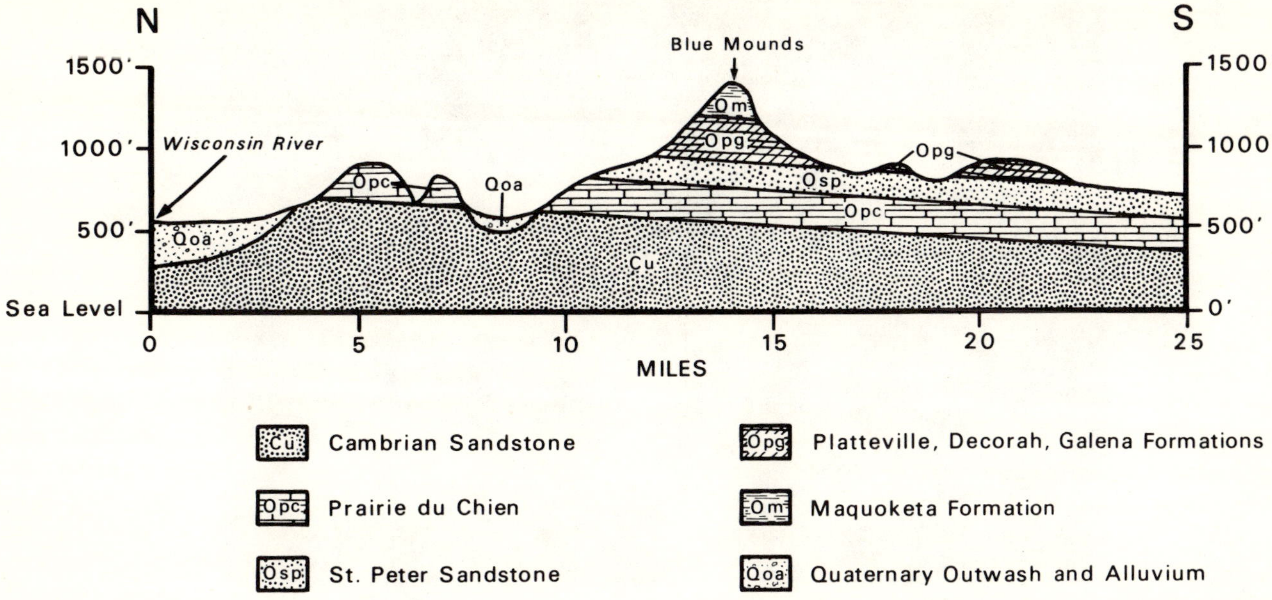

Figure 3: Cross section from Wisconsin River south across East Blue Mounds to the southern edge of the county. West Blue Mounds, which is capped by Niagaran dolomite, is just to the west, outside Dane County. The whole section is underlain by Precambrian igneous and metamorphic rocks.

Eventually, a reduction in the sand supply to the ancient sea shifted the rock type being formed from primarily sandstone to dolomite or limestone, a deposit of magnesium/calcium carbonate similar to that being deposited in Florida or the Bahamas today. This took place early in the Ordovician Period, about 460 million years ago, and resulted in the Prairie du Chien dolomite, which is exposed in Black Earth Creek valley and in other parts of the county. Dolomite can be distinguished from sandstone by its blocky appearance in outcrop (compare Figures 4 and 5).

After the sea retreated, erosion cut stream valleys across the dolomite and sandstone surface and southern Wisconsin was then again inundated by the sea that deposited the St. Peter sandstone. The very pure quartz sand of this rock unit was probably once part of the Cambrian sandstones before being eroded and redeposited by rivers and waves. Outcrops of St. Peter sandstone are common in western and southern Dane County, where the color varies from white (or gray) to rust where it is iron stained. An example can be seen in Figure 5 and several localities are mentioned in the self-guided fieldtrips at the back of this book.

As happened at the end of Cambrian time, the ocean environment eventually changed, and dolomite and limestone were deposited to form the Galena and Platteville Formations (Figure 2) which overlie the St. Peter. Exposures of those units are also common in southwestern Dane County.

Shale was deposited next (Maquoketa Formation, Figure 2) but this has been removed by erosion from most of the county. Even where present, at the top of East Blue Mound and on the east shoulder of West Blue Mound it is poorly exposed. Just west of the county line, atop West Blue Mound, another dolomite unit can be seen that was deposited during Silurian time. This is the youngest rock in southern Wisconsin (420 to 435 million years) and it is exposed only in the highest

Figure 4: The Galena dolomite exposed in a road cut on County PD, west of Verona Road, southwest of Madison. Note blocky nature of dolomite in outcrop.

areas because erosion has removed it elsewhere. This dolomite is the same rock unit that forms Niagara Falls in New York state. Because the unit is resistant to erosion, it also forms a ridge in eastern Wisconsin from Walworth County northward to the tip of Door County which is popularly called the Niagaran escarpment.

The bedrock has an influence on the appearance of landscapes in the county. The rock layers dip or tilt toward the south and the deepening of the Wisconsin River valley has provided an asymmetric shape to the land surface in western Dane County (Figure 3). Military Ridge, the east-west highland that runs from the Madison vicinity west through Blue Mounds and eventually to the Mississippi River, is the divide between the Wisconsin River to the north and Sugar and Pecatonica Rivers to the south. To the north of this divide relatively short, steep, and deeply incised streams flow to the Wisconsin River. To the south of Military Ridge, valleys are broader and more gently sloping as their streams carry water and sediment to the south.

Figure 5: St. Peter sandstone exposed in a roadcut on Mineral Point Road west of Madison. Contrast the appearance of this sandstone with the dolomite in Figure 4.

This asymmetry can be seen from the towers atop Blue Mounds. The valleys to the north are more wooded than the ones to the south, reflecting the difference in steepness that causes the difficulty in farming the land. This difference can also be seen in the character of the streams themselves. South of Blue Mounds the streams have more gentle gradients and carry sand along their beds. To the north, in the steeper valleys, streams carry pebbles and cobbles of dolomite and chert (very hard silica deposits) derived from the dolomite.

Since deposition of the Silurian dolomite some 400 million years ago, the geologic record of the area is unclear because erosion was dominant. There are no known deposits in Dane County dating from Silurian time to Late Quaternary time, less than 100,000 years ago, although marine deposits between 63 and 135 million years old are present in western Minnesota. In places, reddish-brown clay that is the remnant of weathered dolomite is present beneath the modern soil.

Processes of slopewash and slow downslope movement (called creep) which move material down the valley sides, are active in the Driftless Area. On slopes along Black Earth Creek, surface soils are moving, on the average, about 5 mm per year in the downslope direction.[9] Although this rate seems small, over millions of years huge volumes of material are moved from the slopes into the valley bottoms where they are then transported out of the country by streams. This process, along with surface wash from rainfall, continuously lowers the land surface through time.

One of the more striking features of the Driftless Area is the presence of sandstone monuments or towers that stand above the surrounding countryside. Several of these are visible in the area around Mt. Vernon and eastward to Footville (which is actually in the area covered by older drift). These monuments formed recently in geologic time, probably under colder, wetter conditions during glacial time. It is possible that differential erosion created these monuments within the last 20 to 30 thousand years.[10] Within the area glaciated more recently, monuments are small and not nearly as spectacular. The reason for the monuments is not completely known, but an important factor is that parts of the sandstone are more resistant to weathering and erosion because they were differentially cemented by fluids moving through the rock. Similar monuments occur throughout the state where sandstone is the surface bedrock. One of these, Mill Bluff in Monroe County, is a state park and will become part of the Ice Age National Scientific Reserve. Several other parts of the state also have these sandstone pinnacles present, and many can be seen along I-94 between Wisconsin Dells and Eau Claire.

The size of grains being carried by the streams changed drastically after this. Most of the fine-grained sediments (silt and clay, or grains smaller than 0.0625 mm) that we see in valleys today were deposited during the Holocene (the last 10,000 years). As glaciers retreated from Iowa and Minnesota, and exposed an unvegetated ground surface, huge volumes of loess (primarily silt sized [.0625 - .02 mm] dust carried and deposited by wind) were deposited east of the Mississippi River. Much of the loess probably was derived from the broad bottomland itself as water levels fluctuated daily and seasonally, exposing fresh material to drying and then to wind erosion. Because even then the prevailing winds were westerly, the loess forms a blanket, thickest (over 10 m locally) on the west edge of Wisconsin and thinning to the east. In Dane County, thickness is usually less than 1 m. This material was, and is, easily removed from slopes by wind and water.

During parts of the Quaternary, when glacial ice was present north and east of the Driftless Area, downslope movement of materials was probably much more rapid than it is today. At that time, tundra vegetation covered the uplands, spruce forests grew in the valleys, rainfall was probably somewhat higher, and permafrost (permanently frozen ground) existed throughout much of the area.[11] Because of the permafrost, upper parts of the ground became very wet during periods of summer melt, causing a great deal of downslope movement of soil and bedrock. During times of lower temperature, streams in the Driftless Area may have been quite different than they are today.[12] Because of greater amounts of rainfall and because of the huge mass of material that was being added to the streams, the streams were probably broader than they are now and consisted of numerous channels flowing over a gravel bed. This gravel bed can be found beneath the finer-grained alluvium or stream deposits throughout much of the Driftless Area.[13]

Much of the loess was probably dropped on a vegetated land surface. In far western Dane County, just north of Blue Mounds, Davis[14] examined the pollen that is preserved in peat on top of the gravels and below the fine-grained sediments that can be seen in all of the floodplains in this area. In this peat, which contains logs, snail shells, and fibrous plant remains, the high percentage of spruce pollen near the base indicates that spruce was the dominant vegetation in this area (at least in the valleys) until about 9,500 year ago. During this time of abundant spruce pollen, the earliest documented record of human habitation occurs in southern Wisconsin. Several sites, including ones in the glaciated area, have produced fluted points, artifacts characteristic of cultures living between 10,000 and 11,500 years ago. Throughout Dane County at this time, mastodons and perhaps mammoths, both types of elephants, roamed the open spruce or boreal forest.

Three well-documented mastodon sites exist in eastern Dane County near Deerfield and scattered remains have been found elsewhere. All of the finds are in peat, and there are probably many more remains scattered through the wetlands of Dane County.[15] Giant beaver (up to 500 pounds in weight and 7 feet long!) probably coexisted with mastodons in this environment.[16] About 10,000 years ago, a rapid decrease in the amount of spruce accompanied a major increase in pine and oak vegetation.[17]

Since that time, pine and oak pollen have characterized southern Wisconsin. Studies of pollen in Holocene peat deposits reveal fluctuations in pine, oak, and types of pollen such as grasses that suggest a considerable change in climate during the last 10,000 years. Between 9,000 and 5,000 years ago, the climate was evidently warmer and dryer than today and open jack pine and oak woodland bordered the prairie to the west. Today, there are few archaeological remains to record the activities of inhabitants at that time. Mastodon and giant beaver populations had probably declined by 9,000 years ago and deer and elk became more common. After about 5,000 years ago, when deciduous woodlands somewhat similar to those of today began to predominate, the archaeological record becomes much more complete and complex.

Evidence of this Holocene climate change can also be seen in the record preserved by stream sediments. About 10,000 years ago, streams changed from their braided, shallow condition with numerous channels into the single, narrower channel form they have today.[18] Since that time, the floodplains that we see along present-day streams have been alternately built up vertically by accumulations of silt and sand deposited at times of flood, and eroded by the migration of stream channels from one side of the floodplain to the other. The meanders (Figure 6) that we see in many of the streams in Dane County do not remain in

Figure 6: Meandering stream typical of many of the smaller streams in the county. Note that the steep bank is on the outside of the meander where active erosion is taking place. Abandoned meanders (oxbows) showing older positions of the channel are present on the floodplain surface. Photo by J. C. Knox.

Figure 7: Post-settlement alluvium deposited during the last 100 years over old soil which represents the former surface of the floodplain. Photo was taken of stream bank in the Driftless Area.

one place through geologic time but migrate slowly but surely downstream and often from one valley wall to another. This migration is produced by erosion on the outside banks of meander bends and deposition on the inside banks of meander bends (Figure 6). On many floodplain surfaces crescent-shaped depressions, often filled with organic sediment or water, mark the location of earlier channel positions.

The most recent change of streams, especially those in the Driftless Area, came with the advent of land clearing in the 1800's. In many places much of the fine grained alluvium above the underlying sand or gravel surface (Figure 7) was deposited in post-settlement time.[19] Very often this "postsettlement alluvium" is several feet thick and covers an old soil which represents the former floodplain surface. Old pots, bottles, fence posts, and other signs of early habitation and agricultural development in western Dane County have been found on the buried soil beneath this postsettlement alluvium. The change in the deposition rate of the streams took place because of land-clearing practices. Vegetated slopes were cleared of forest vegetation and plowed, often with little thought toward preservation of the soil, and a great deal of erosion took place on the hill slopes.

THE GLACIATED AREA

As stated in the Introduction, glaciation has had a profound influence on the landscape and soils of Dane County. Although some older glacial deposits are present, by far the most important glaciation modifying our landscape was the Woodfordian or late Wisconsin glaciation, which took place 15,000 to 13,000 years ago. The maximum extent of that glacier is marked by the Johnstown Moraine, which was deposited at the glacier terminus (Figure 1 and map in back pocket). The Johnstown Moraine enters the county just east of Brooklyn and north of Evansville and trends northwestward through Verona to Cross Plains. From Cross Plains to the northwest corner of the county, the glacier ice was relatively clean and left no ridge at its terminus.

As the ice retreated from this moraine, it experienced a minor pause and deposited the Milton Moraine, a recessional moraine that passes just west of Stoughton and also trends northwestward, crossing Black Earth Creek between Middleton and Cross Plains (Figure 1 and map in back pocket). This moraine trends to the northwest as a discontinuous ridge and leaves the county north of Roxbury. To the east of these moraines the Yahara Lowlands, formerly a valley cut in the bedrock by preglacial streams, contained a fairly large lake as the ice retreated. To the east and north of this lowland, the prevailing glacial landforms are gently rolling plains and drumlins, which formed beneath fairly thick glacial ice, while the terminus of the glacier was out at the moraines.

The four landform zones within the glaciated area will be discussed separately. I will start with the westernmost zone of older glacial drift, then discuss the end moraine zone, the Yahara Lowland, and finally the ground moraine zone with drumlins in the eastern part of the county.

Before this is discussed, however, we should look at the types of materials and landforms that are left by a retreating glacier. There are two main types of sediment: stratified drift and unstratified drift. Unstratified drift refers to the material that is deposited directly by glacier ice. It shows no stratification, or layering, that is typical of water deposited sediments (Figure 8). The term that is used for these un-stratified materials is till. Till in Dane County is usually composed of a matrix of sand, silt, and clay (average of 61%, 29%, 10%, respectively), with a fairly large number of pebbles and larger rocks up to boulder size. Thus a large range in grain size is present in till. Till is a common material in moraines and drumlins and is the surface material in areas mapped ground moraine (Figure 9 and map in back pocket).

The other category of drift is stratified drift (Figure 8). Stratification, or layering, in the deposits implies deposition by water. Two major types of stratified drift are (1) outwash, or material that is carried by streams washing off or out of the glacier, and (2) ice-contact stratified drift, which is deposited in contact with glacier ice (Figure 9). Both of these types of deposits show layering typical of water deposition. However, outwash typically is much more regular and predictable in terms of the size of materials it contains and usually it has a relatively flat surface.

Figure 8: Till overlying sand and gravel. Note the differences between the well-mixed, unstratified, and poorly sorted till and the layered or stratified sand and gravel. Water deposition produces the layering and also segregates materials of different sizes.

In places where outwash has been deposited on top of melting ice blocks separated from the main glacier, pitted outwash, or outwash with closed depressions (kettles) forms as the ice blocks later melt out (Figure 9). This type of outwash grades into ice-contact stratified drift as the amount of ice left beneath the sediment increases. If the sand and gravel was deposited over a continuous, fairly thick mass of ice at the glacier margin, the melting of the glacier ice at a later time produces a very hummocky landscape underlain by ice-contact stratified drift. This landscape is what is called kame and kettle (or knob and swale) topography with the term <u>kame</u> referring to the hills and the term <u>kettle</u> referring to the closed depressions (Figure 9).

Because much less water transport takes place in the deposition of ice-contact stratified drift, this material usually is more poorly sorted (that is, it has a larger range in grain size) than outwash. Also, it usually is very variable, and sand and gravel operators working in this type of deposit find it difficult to predict how coarse the materials will be from week to week as they work in a gravel pit. The use of these materials as aggregate is discussed in a later section.

Older Drift Zone

This small part of the county resembles the Driftless Area with respect to its landscape, although glacial deposits are present on much of the land surface and a few very subdued glacial features can be seen just west and south of Brooklyn. The age of the drift in this region is not known, but it probably predates the major glaciation that covered the remainder of the county.

It has been speculated that the area shown on the map, Glacial Geology of Dane County, Wisconsin, as pre-Woodfordian dates from the early Wisconsin glaciation or is somewhat older

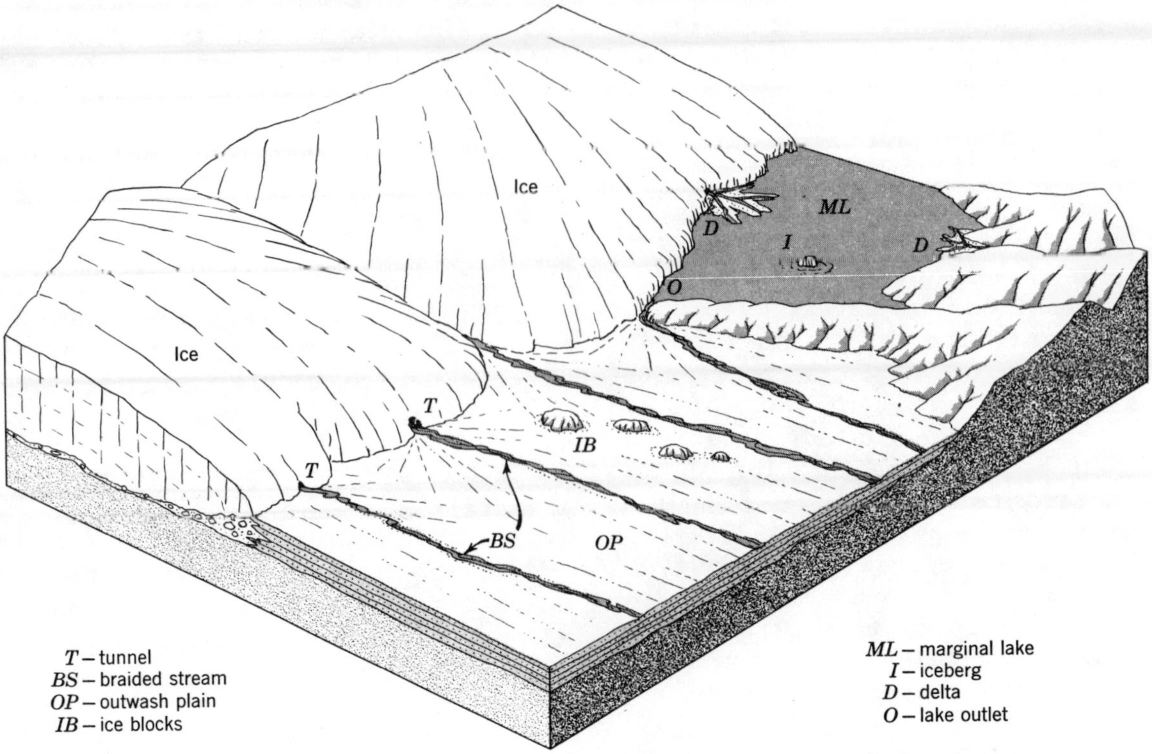

A. With the ice front stabilized and the ice in a wasting, stagnant condition, various depositional features are built by meltwater.

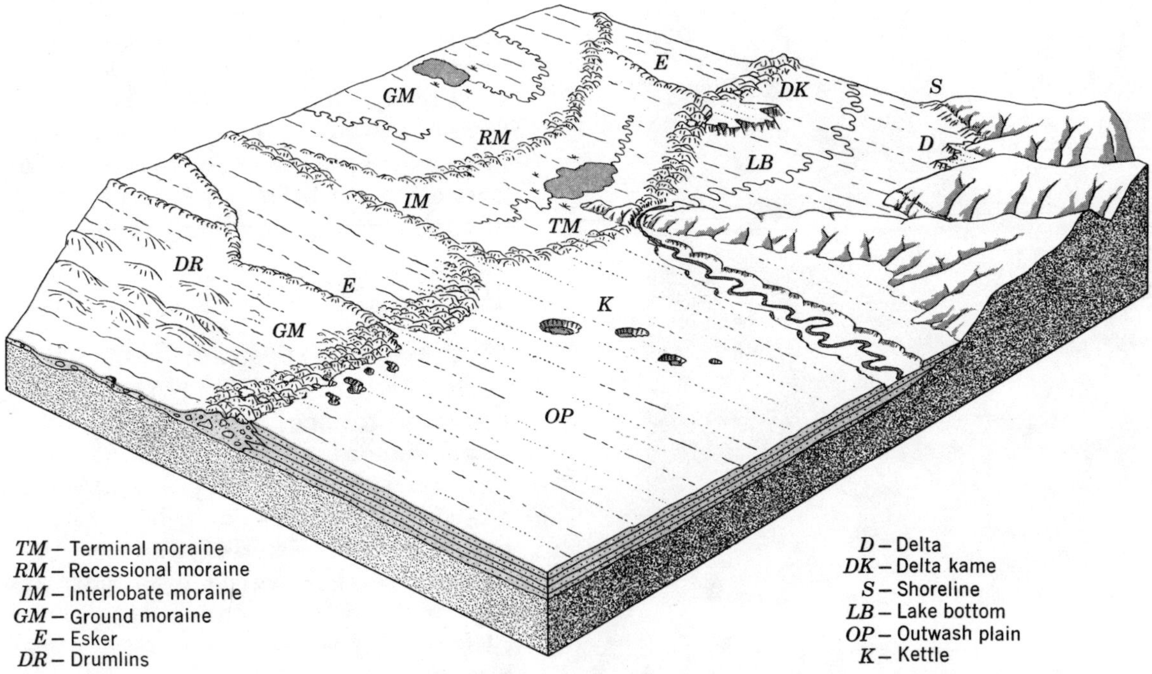

B. After the ice has wasted completely away, a variety of new landforms made under the ice is exposed to view.

Figure 9: Block diagrams showing typical landforms developed by continental glaciation. Taken from Strahler (1960, Physical Geography, New York, John Wiley and Sons).

than 30,000 years.[20] In this area the thickness of loess, or windblown silt, is greater than it is on the younger glaciated surface and there are exposures of bedrock in many of the steep slopes. Soils that are mapped on this surface are also deeper than soils on the younger glacial surface and this suggests greater age.

Between this area and the Johnstown Moraine, or the area mapped as early Woodfordian (egm and eem on the glacial map) are somewhat younger deposits, which resemble the even younger deposits in the Johnstown Moraine adjacent to the north and east. Very little of this drift is exposed and its relationship to the adjacent deposits in the Johnstown Moraine is not clear. These deposits may represent a glaciation that took place perhaps several thousand years earlier than the ice advance that came to the Johnstown Moraine.[21] Based on the soils that are developed on this drift, it is unlikely that the deposits are much older than 20,000 years but their position in front of the Johnstown Moraine makes them older than the moraine itself. Another possiblility is that in this area ice that was at the Johnstown Moraine in other parts of the county and to the south in Rock County, extended somewhat further here, leaving the so-called Brooklyn Moraine before retreating to the position of the Johnstown Moraine a few miles to the north. Because there is so little difference in age and no datable materials have been found, we cannot determine the exact age relationship between these moraines.

End Moraine Zone

In a sense, glaciers can be thought of as giant conveyor belts moving material from thicker ice to thinner ice near the margin. Thus, the drift of Dane County is made up of local bedrock material and also some material that has been carried from parts of northeast Wisconsin and Canada. Since glacier ice flows plastically toward the margin, this material, incorporated somewhere up-ice, is moved toward the margin, where it is deposited. Glacier ice always flows toward the ice margin so the operation of the "conveyor belt" is continuous, even when the ice margin position retreats because of increased melting or lowering rates of flow.

Debris can be deposited at the base of the ice (basal till) or melted out on the surface and let down as underlying ice melts (supraglacial till). Nearly all of the till in Dane County is basal till although in places, especially in the moraines, a thin layer of supraglacial till is present. This material is looser, sandier, and more rich in dark-colored igneous rocks than the basal till.

If a glacier advances and then retreats at a relatively constant rate with no pauses, a veneer of till of more or less equal thickness is deposited across the landscape. If, however, there are pauses in the retreat or advance of the margin so that the margin sits in one place for some hundreds of years, then deposition of till and associated ice-contact deposits builds a ridge parallel to the ice edge that is called an end moraine (Figure 9). The name terminal moraine is used for the outermost end moraine of any given advance. The other end moraines behind or up-ice from the terminal moraine are called recessional moraines because they are deposited during the recession of the ice margin. Some moraines probably are also built as the ice advances, but very often these are destroyed by the overriding ice.

The end moraine zone in Dane County is made up of two principal moraines; the outer, terminal moraine is called the Johnstown Moraine and the inner, recessional moraine is called the Milton Moraine. About 14,000 years ago, ice sat at the position of the Johnstown Moraine (Figure 10). North of Cross Plains no terminal moraine was built probably because the ice was

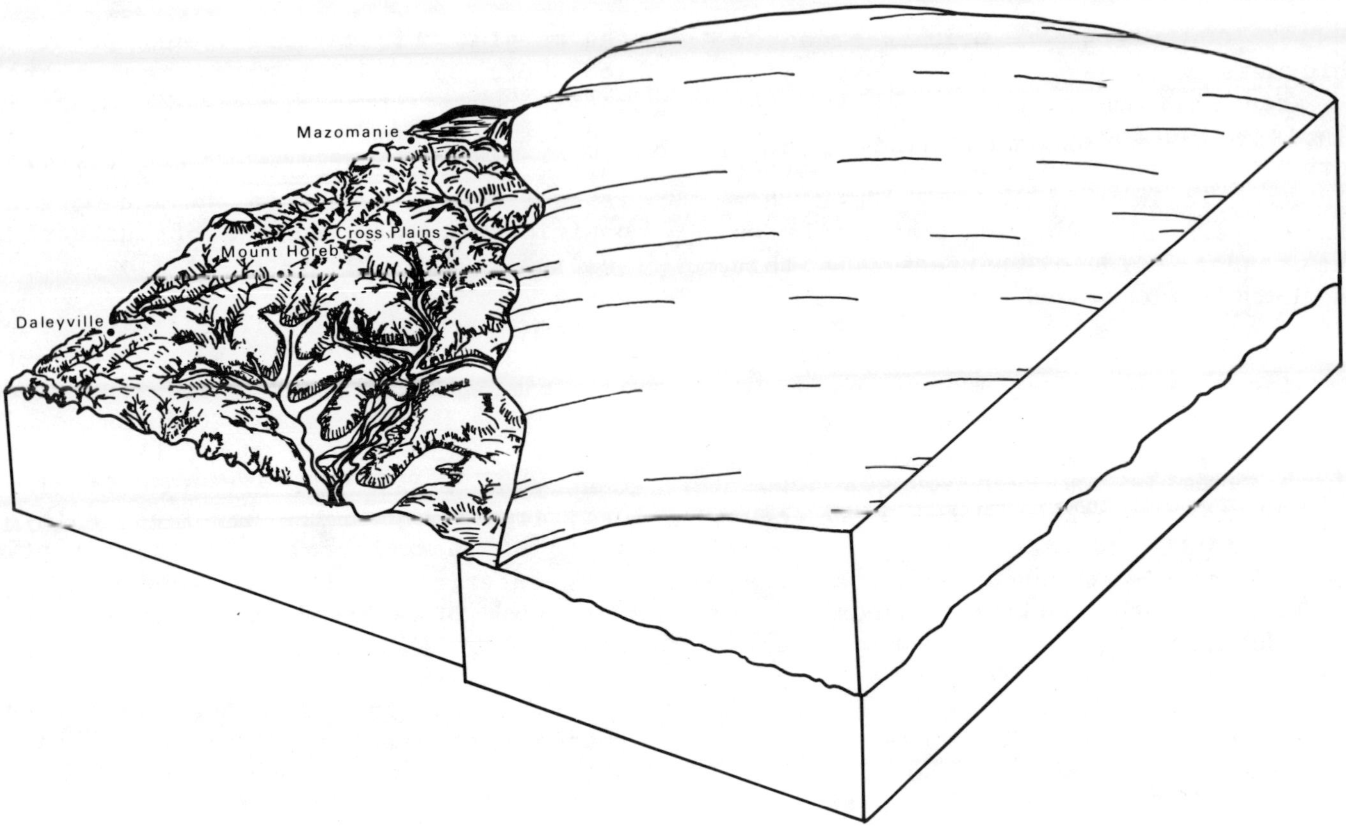

Figure 10: Landscape diagram showing the position of the ice at the time the Johnstown Moraine formed, probably about 14,000 years ago.

Figure 11: Air view of the Johnstown Moraine in foreground and outwash plain in background. Note the difference in topography and land use between the two landforms.

relatively clean and little till was deposited. A thin scattering of till does, however, occur out to the location of the glacier margin. In addition to the Milton Moraine, a few minor recessional moraines also occur in the eastern part of the county but these are very small and are difficult to detect without a detailed topographic map and investigation in the field.

Although the end moraines in Dane County are not very high, they do have a characteristic topography which makes them recognizable to the practiced eye (Figure 11). Very often, the moraines have a scattering of fairly large boulders which, if they have been cleared from the fields, occur as piles or small stone walls on the edge of cropland. Small gravel pits in the moraine also show this very bouldery till and ice-contact stratified drift. The moraines usually have kettles or closed depressions that are developed as blocks of melting ice disappear. These often now contain marshes or ponds because of the poor drainage. Near the crest of the moraines, the landscape often contains kettles and very steep slopes along with the bouldery material that makes farming difficult. For this reason, many of the moraines in Dane County are wooded, at least along their crests. The pattern of moraines that can be seen on the satellite photogaph (cover) is partially due to this difference in vegetation.

While glacier ice was at the terminal moraine about 14,000 years ago, large volumes of meltwater were being produced and carried away from the glacier margin. The three main outlets of glacier meltwater were the Sugar River to the south, Black Earth Creek and some of its tributaries and the Wisconsin River and some of its smaller tributaries in the northwest part of the county.

As meltwater flows from a glacier margin, it carries coarse debris along with sand, silt, and clay and deposits these materials in the valleys away from the ice margin (Figure 9). This outwash shows features typical of water deposited material in that it is layered, or stratified, and is relatively well sorted (material of various sizes is separated both vertically and horizontally). Thus, if we examine the size of materials in the outwash, we find that the coarsest material is adjacent to the moraines, and that materials become systematically finer downstream. This can be seen along the Sugar River, where gravel pits right at the edge of the moraine just outside of Verona have material up to one foot in diameter. The outwash becomes finer downstream, and at Belleville, the outwash consists mostly of sand with a few small pebbles.

For this reason, many gravel pits throughout Wisconsin are in outwash close to moraines. The location of a number of these gravel pits is pointed out in the road log for the field trips. The relationship between deposits in the moraine and adjacent outwash is not well exposed in the vertical sections of gravel pits in Dane County. Figure 12 shows a photograph from a gravel pit along the moraine in Sauk County just east of Devils Lake. Note that the material to the right in the photograph, in the moraine, contains very large boulders and very little stratification or layering. To the left, or west of the moraine, former outwash streams have deposited layered outwash under the sloping surface.

Land use differences between the moraine and the adjacent outwash are readily apparent in many places in the county and to the south in Rock County. Because the moraine is often wooded or pasture because of its rough surface and bouldery materials, it is very distinct from the finer materials and smooth surface of the outwash plain. The outwash is well drained, but because it has a covering of loess, it has enough moisture retention to make it prime agricultural land. These relatively flat surfaces of the outwash

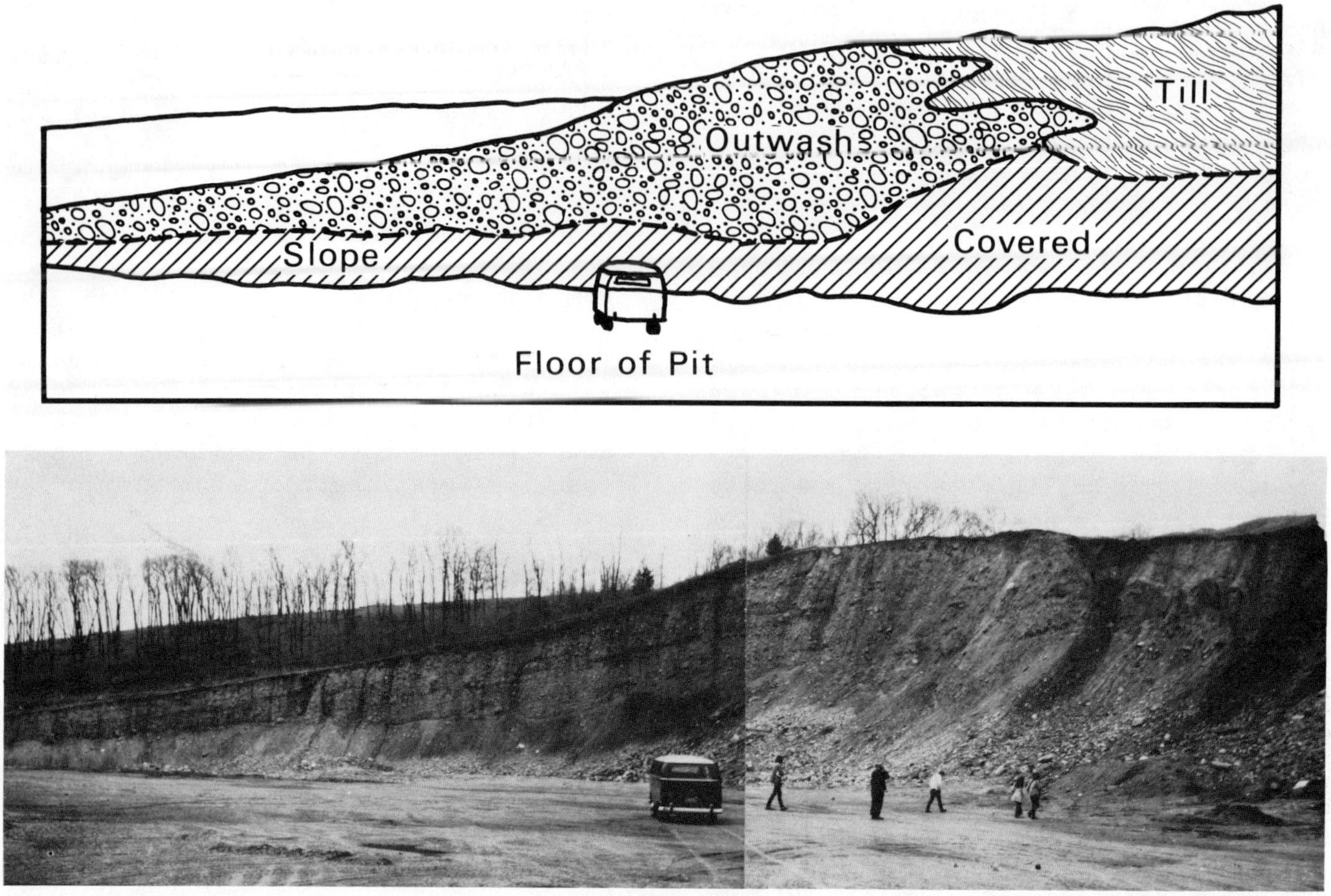

Figure 12: Gravel pit in the moraine edge in Sauk County east of Devil's Lake. Note that to the right till is exposed in the crest of the moraine. Large boulders can be seen in this material in the slope and also at the base. To the left, away from the crest of the moraine, is outwash. Note the layering or stratification in the deposit and also the lack of large boulders which are present in the adjacent till. When this material was being deposited, ice was sitting along the crest of the moraine to the right and meltwater was flowing from the ice downhill toward the left.

provide an excellent substrate for row crops such as corn.

Some smaller drainage features are also present along the moraine. In some places, instead of depositing outwash, streams of meltwater cut channels through the bedrock. One place where this has occurred is on the south wall of the valley of Black Earth Creek between Cross Plains and Middleton at Wilke Gorge. This area, which will become part of the Ice Age National Scientific Reserve[22] contains a gorge (Figures 13 and 14) which was cut down along and under the ice margin while it was at the Johnstown Moraine. Because glacier ice blocked some of the preexisting valleys, a small lake and ice-marginal channel developed along the front of the moraine from Mineral Point Road north to the site (Figure 14). The lake drained along the ice margin and finally under the ice and into Black Earth Creek valley, cutting the gorge which exists today.

While the ice was at the terminal moraine, some melting was taking place beneath the ice sheet to the east. This water was collected in subglacial drainageways and was carried underneath the glacier in tunnels toward the ice margin. In some cases, valleys were cut in the landscape beneath the ice and in some places sand and gravel was deposited underneath the ice (Figure 9). Very often, when this happens on a small scale, a single, distinct ridge of sand and gravel, called an esker, forms (Figure 9). In Dane County, however, the tunnels were evidently fairly large and there are no well-developed long eskers, as are present in eastern and northern Wisconsin. Instead we have groups of kames and kettles in sand and gravel deposited in parts of these large subglacial valleys. These deposits are mapped as ice-contact stratified drift on the map, Glacial Geology of Dane County, Wisconsin.

The best developed subglacial drainage system in the county is one in the northern part, extending as groups of kames and kettles near Norway Grove, southwestward through two more-or-less parallel valleys just north and west of Waunakee, westward through Waunakee Marsh and a valley to the north of it, to another ice-contact deposit near Springfield Corners (see the Glacial Geology map of Dane County, Wisconsin). Highway 19, west of Springfield Corners, follows the valley that ends abruptly about one mile west of Highway 12. This sub-ice drainage was evidently operating when ice was at the Milton Moraine, and large volumes of water must have been draining throughout the subglacial channel.

Other similar subglacial channels occur in eastern Dane County. One of these is the valley running parallel to, and just southeast of, Highway 151 near Sun Prairie. Two other major deposits of ice-contact stratified drift that probably developed subglacially are present in the southeast corner of the county near Lake Kegonsa and just northeast of I-90, north of the town of Utica and west of Rockdale.

Two more of these deposits occur in the southern part of the county just south and east of Oregon. Another similar one, evidently also related to a drainage system, is present in the Five Points area at the intersection of Highway PD and M. This subglacial channel, which leads from a valley to the east, evidently contributed outwash to the flat plain present just in front of the Johnstown Moraine on Highway PD.

While very little detailed study has been done on these subglacial channels, they are present throughout the Green Bay Lobe and can be found just north of Janesville and also along the terminal moraine north to Antigo in Langlade County.

As glacier ice retreated from the Johnstown Moraine, outwash was deposited over isolated ice blocks in places between the moraines. Where the source of outwash was cut off before the ice blocks melted, pitted outwash (Figure 9) was produced because no later outwash filled the depressions. This type of landscape can be seen between the moraines southeastward from Oregon. When the ice reached the Milton Moraine, much of the drainage was concentrated in Black Earth Creek from Cross Plains westward, down the valley from the Milton Moraine to Verona along Highway 18 and 151, and several other smaller valleys. Several small and short lived lakes were also formed at this time, but they are of minor significance compared to the lakes to the east.

Yahara Lowlands

As the ice melted back from the Milton Moraine the elevation of the ice margin became lower than the land surface elevations in the moraine zone because of the preexisting Yahara River valley. Major drainage changes took place at that time because many of the tributaries that formerly carried outwash were cut off from their source of meltwater as ice retreated into the lowland. In the Middleton area, because the outwash surface of Black Earth Creek was higher than the land surface to the east a lake formed along the ice margin. This lake is named Glacial Lake Middleton[23] and it drained into Black Earth Creek. The extent of the lake is shown on the Glacial Geology map of Dane County. Much of the sediment deposited in the lake is sand and fine-grained silt and clay, and there are some organic materials along the margin. Old beaches can be seen, especially along the north side of the lake, where on freshly plowed ground they show as a lighter-colored, better-drained soil.

As soon as the glacier ice had retreated far enough for drainage to shift to the southeast (probably near the University of Wisconsin-Madison), water level dropped from Glacial Lake Middleton to a lower level of a much more extensive lake, which is referred to here as Glacial Lake Yahara. Because there are no relevant radiocarbon dates, we do not know how long Glacial Lake Yahara was in existence. It probably was fairly short lived, although its deposits are quite extensive. Deposits of this lake are shown on the Glacial Geology map of Dane County and the extent of the lake is shown in Figure 15. It can be seen that the Four Lakes area was one large lake at one time. The highest elevation of Glacial Lake Yahara was approximately 860 feet, or about 12 feet higher than the present Lake Mendota. As the lake drained it evidently had several lower stages and some shorelines are present at about 852 feet, or about 4 feet higher than the present Lake Mendota.

At the time Glacial Lake Yahara was in existence, most of the water in central and eastern Dane County was

Figure 13: View looking upstream in the gorge which will become part of the Ice Age National Scientific Reserve. Note that the channel is now dry because the cutting was entirely due to glacial metlwater.

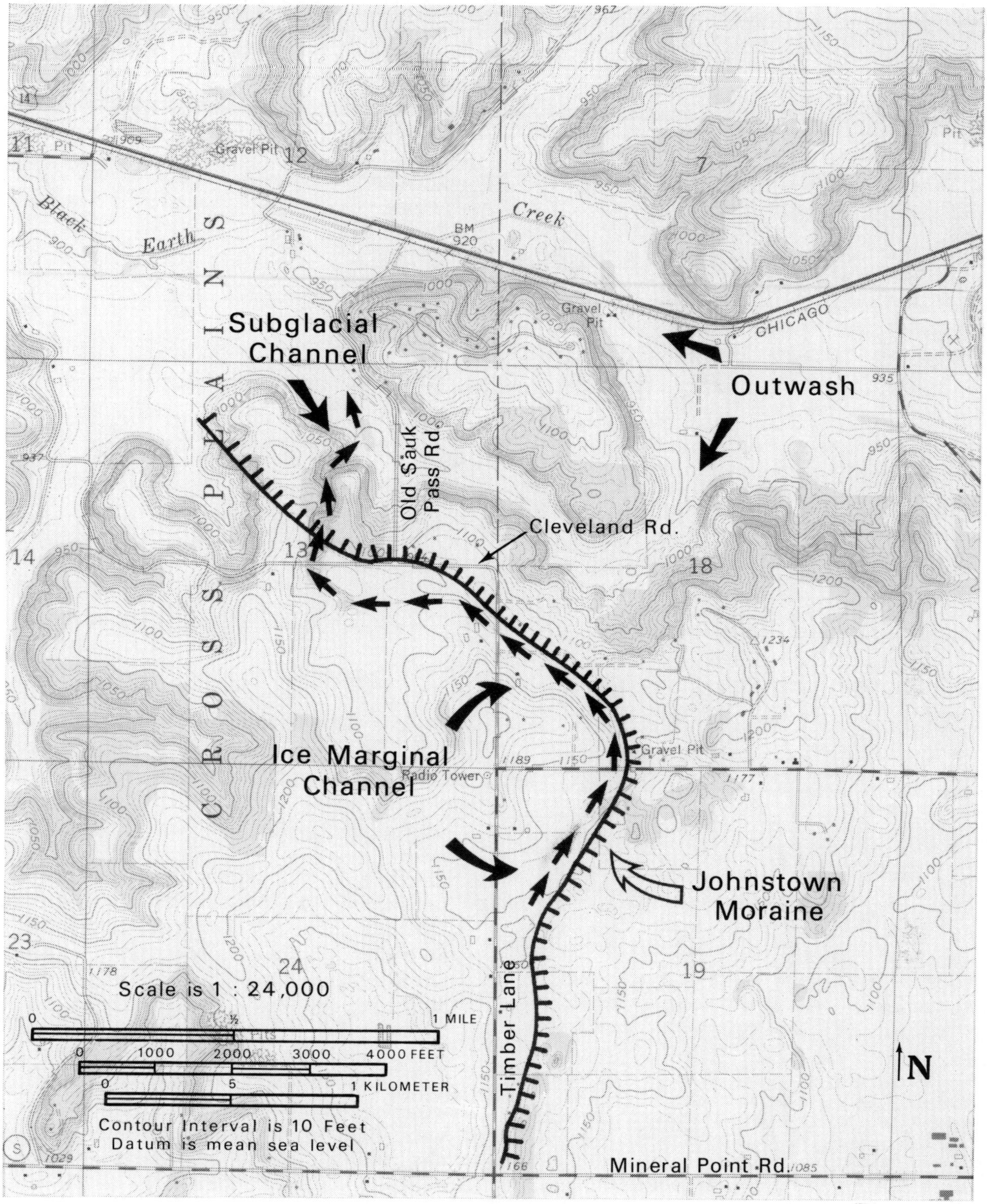

Figure 14: Topographic base map showing glacial geology of the area surrounding the Cross Plains site. Black Earth Creek is in the upper or northern part of the diagram and Mineral Point Road crosses the map east to west near the bottom. Symbols are those used on the glacial geology map of the county.

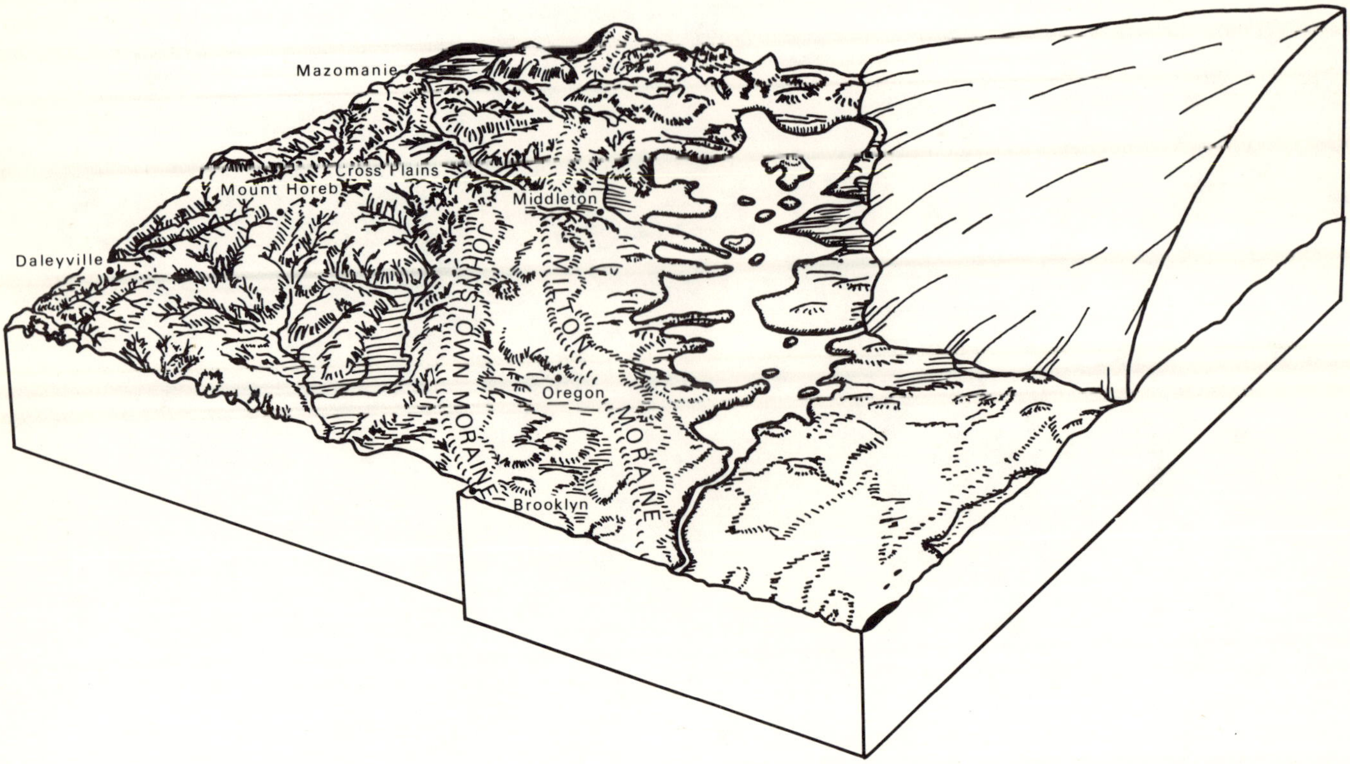

Figure 15: Schematic view of Dane County as ice retreated to produce glacial Lake Yahara.

passing through the lake system. Water was draining from the north through the valley near Lake Windsor and from the melting ice margin to the east down numerous valleys into the lake. Probably early in its history Glacial Lake Yahara was draining over buried glacier ice in the Milton Moraine. Some remnants of outlets can be seen near Stoughton, but the complete history of the lake cannot be reconstructed because melting ice in the Milton Moraine destroyed some of the possible outlets.

Much of the glacial lake plain developed a surface of organic-rich materials (peat) because the soils were poorly drained and very low in the landscape. In Madison itself, and in some of the surrounding suburbs, much of this former lake area has been artificially filled during urbanization. In some places, however, sand is still present at about, or just below, the former elevation of the lake. The flat surface on which Truax Field is built is part of this surface, as is the flat sandy surface near the intersection of I-90 and Highway 12 and 18. Areas east of the Capitol, in the Monona Bay and Lake Wingra area, University Bay, along the west edge of Lake Mendota in Middleton, and Cherokee Marsh are all areas where fairly extensive organic deposits cover the lake sediment.

Beaches from the former lake are not well preserved although they can be seen along the steep slopes in Olin Park (at the south boundary) and along the south edge of Lake Monona. At these sites, just above the level of present Lake Monona, one can see a shoreline at an elevation of about 852 feet and another fainter shoreline at an elevation of about 860 feet. The lower shoreline is also well developed along parts of Picnic Point. Paleoindian, as well as some later artifacts, have been found along the edges of the former lake.

The evidence that Glacial Lake Yahara was one large lake is that lake deposits occur up to an elevation of about 860 feet throughout the Yahara River system south to the Township of Dunkirk, just south of Stoughton. As the lake was probably short lived, the shoreline features at 860 feet are not very well developed and most have been destroyed by subsequent erosion. Just southeast of Stoughton, however, there is a remnant of the old lake plain between elevations of 850 and 860 feet. One possible outlet of the lake can be seen along the east side of this lake plain (Figure 16). This outlet would have been active while glacier ice was buried in the Milton Moraine, damming the present valley of the Yahara River near Dunkirk.

Although there is evidence that glacier ice can remain buried under debris in moraines for thousands of years[24] we have no evidence of the length of time that ice was present in the Milton Moraine. As ice melted out of the moraine, the Yahara River probably switched its course to its present one and began cutting through the till and underlying dolomite. For the last 10,000 years or so, the river has cut slowly down through bedrock, which is exposed in the section of the river just north and south of Dunkirk. We can hypothesize that the level of the lake dropped fairly rapidly from the 860-foot level to an elevation of about 852 feet, and since then water levels have dropped slowly, especially in the lower end of the Yahara River, where the water elevation is about 830 feet at present.

Thus, the present configuration of the Dane County lakes has evolved slowly through the last 8 to 10 thousand years as the outlet near Dunkirk has slowly downcut through bedrock, and organic materials in marshes have encroached and partially filled-in open bodies of water and exposed lake sediment. The present lakes lie in lower areas of the Yahara Lowland where glacial debris did not entirely fill the valley. Eventually, over geologic time they will slowly fill and lower their outlets until they disappear.

Ground Moraine Zone

The general term ground moraine is used to designate all of the till-covered surfaces that are not in linear trends parallel to the ice margin (end moraines). This ground moraine is very extensive in the northern and eastern parts of the county and is interrupted only by drainageways of outwash with a cover of organic materials or the deposits of ice-contact stratified drift associated with sub-ice drainage while the ice margin was far to the west.

The ground moraine landscape is dominated by the streamlined topography associated with elongate hills called drumlins (Figure 1 and the Glacial Geology map of Dane County). The drumlins produce this pattern on the landscape because of their influence on the position of roads and on agricultural patterns. Very often, the drumlins have fairly steep sides and are used for pasture or are left in a wooded condition. In addition, because they stand above the adjacent interdrumlin areas, they have better drained soils (plowed fields on drumlins have lighter tones than the adjacent, more-organic-rich soils).

There are several hundred drumlins in Dane County and this is only a small fraction of the approximately 5,000 drumlins[25] that are found in the drumlin field of the Green Bay Lobe.

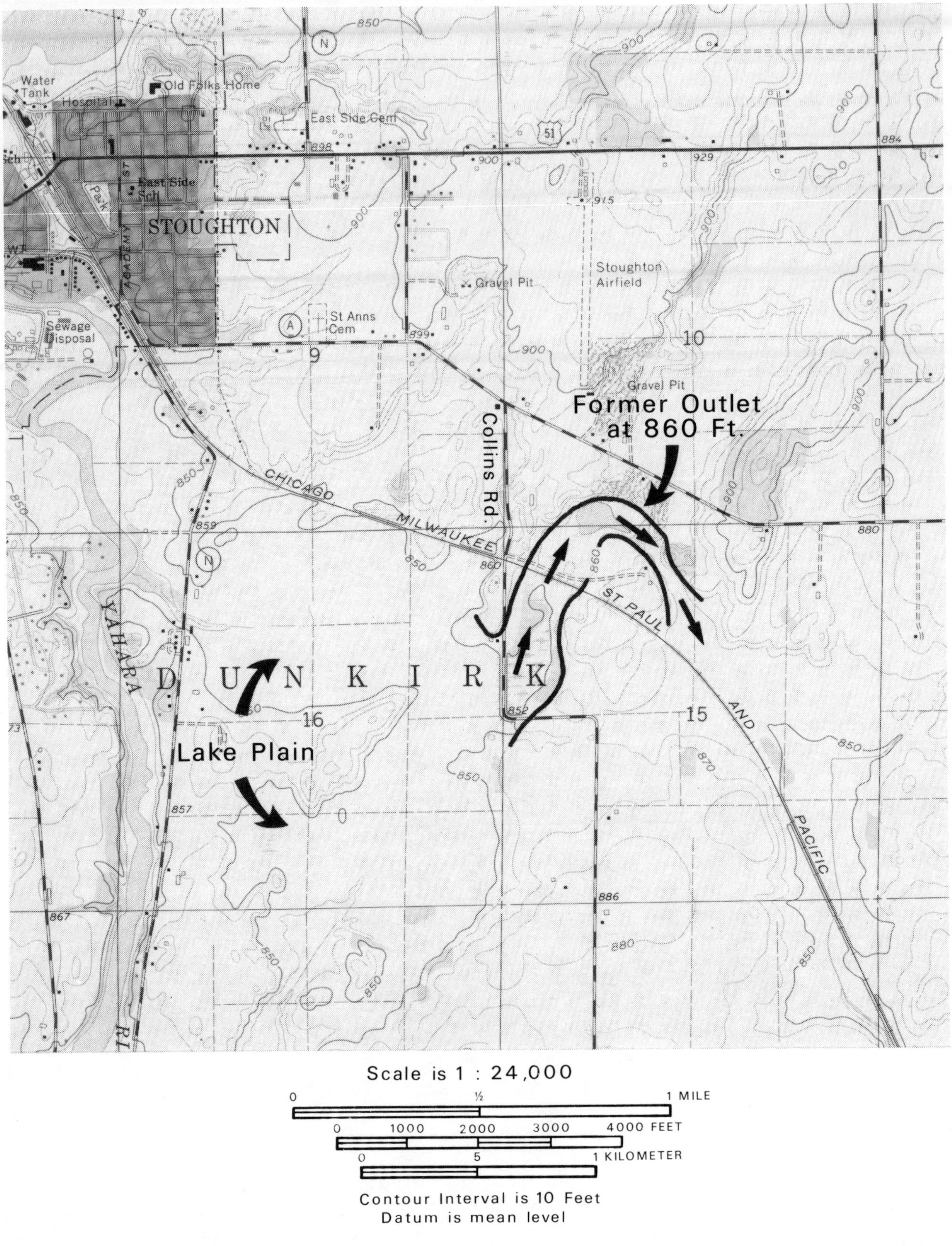

Figure 16: Topographic base and glacial geology of the former outlet of glacial Lake Yahara. City of Stoughton is in the upper left or northwest part of the map.

The landscape that is crossed driving from Madison toward Milwaukee on I-94 or on Highway 151 to Beaver Dam is dominated by drumlins.

The composition of drumlins is variable. Some drumlins are composed entirely of till, some have bedrock cores, and some are composed of sand and gravel with only a thin carapace of till over the top. In the Waukesha area in eastern Wisconsin, many of the drumlins are sources of sand and gravel for concrete aggregate. There are no major sand and gravel operations in the drumlins in Dane County, however, and the composition of most drumlin interiors is not known. Drill records show that the drumlin upon which the state capitol rests and the one forming Bascom Hill on the University of Wisconsin-Madison campus are at least partially cored with sand and gravel. In the late 1800's, Upham noted this fact and wrote a scientific paper describing the "Madison-type" drumlins.[26]

Drumlins have been studied by numerous people for over a hundred years. The streamlining that forms drumlins takes place under fairly thick ice, therefore it is impossible to actually observe them forming. Two main hypotheses suggest that they form either by the deposition of till beneath the glacier ice or by erosion of surrounding material leaving the drumlin as an erosional remnant. Both mechanisms would produce a streamlined form, but the depositional mechanism would also require that the drumlins be made entirely of till. Because a number of drumlins in Wisconsin are composed of sand and gravel with only a thin veneer of till, it seems likely that many of these drumlins formed as glacier ice overrode older glacial deposits and outwash deposits produced in front of the advancing ice. The bed of the glacier produced the streamlined landscape by differentially eroding these materials as it slid over the surface. As the ice retreated, till was deposited in the marginal zone across the landscape producing a nearly continuous till veneer.[27]

We can use studies that have been done on the Antarctic and Greenland ice sheets to reconstruct how thick the glacier ice was at the time the drumlins formed. If the glacier in Dane County had a similar ice-surface profile, when the ice margin was at the Johnstown Moraine, the ice was more than 600 meters (2,000 feet) thick over downtown Madison and in excess of 1,100 meters (or about two-thirds of a mile) thick at the northeast corner of Dane County! It was under these very great thicknesses of ice that the drumlins formed.

Because of the fairly gentle slopes, rolling topography, and general lack of really boulderly material, the drumlin area is relatively good agricultural land. Only on steep slopes of some of the drumlin sides are there problems with soil erosion and with using mechanized equipment. Between the drumlins, the till and outwash is covered with silt and clay below orginic soils. Although some of the fine-grained sediments may represent the former presence of actual lakes, they were probably short lived and much of the fine material has simply accumulated as materials washed off the slopes but were not removed by through-flowing streams.

Sand and Gravel Resources

As well as providing smoother land surfaces and good soils for agriculture, glaciation has provided us with another resource that is important to the economy of Dane County. For any construction using asphalt or concrete, aggregate materials such sand and small rocks are an important ingredient. In 1975, in Dane County, nearly 2 million tons of aggregate, worth nearly 3 million dollars, were produced. About half of this came from crushed dolomite or limestone taken from quarries in bedrock and the other half came from sand and gravel in glacial deposits.

Because sand and gravel has a great deal of bulk and a relatively low value per ton, the distance between gravel pits and where gravel will be used is important in determining the final cost of aggregate materials to the consumer.

Although some till is used for fill in construction projects, sand and gravel in outwash, pitted outwash, and ice-contact stratified deposits are by far the most valuable materials. In 1974, about 65 percent of the sand and gravel produced in Dane County was from outwash and pitted outwash deposits and 35 percent was from ice-contat stratified drift.[28] Because of the differences in the characteristics of these two types of deposits each is best suited for a different use.

Outwash tends to be better sorted, so that fine materials have been naturally washed from the sand and gravel. These deposits are relatively uniform. Therefore, a relatively large capital investment can be made in a mining operation, and very often permanent mining operations are set up in outwash deposits. Due to these characteristics, 81 percent of the aggregate from outwash is used in concrete and only 2 percent in bituminous paving in Dane Couny. Because aggregate that is used for various types of bituminous paving should have more fine-grained silt and clay than is found in outwash, ice-contact stratified deposits are more appropriate. In Dane County 71 percent of this material is used for bituminous paving and only 13 percent for concrete. Because the ice-contact stratified drift deposits are less predictable in their characteristics, they are usually more appropriate for portable-equipment operations. Thus, in many of the smaller gravel pits in the county, equipment is brought in for short periods of time and aggregate is crushed and stored for later use.

The map, Glacial Geology of Dane County, Wisconsin, can be used to evaluate the potential for sand and gravel production for various parts of the county. It should be noted, however, that this only represents potential for development and that the actual value of deposits cannot be determined without on-sight drilling and investigation. In places, conditions such as high water table, thick overburden, location, accessibility, and size of the materials determine the actual value of the deposit. A complete analysis of the sand and gravel industry and potential for mining in the future is given by Hesler.[28]

THE SOIL PATTERN OF DANE COUNTY

Dane County is still one of the leading dairy counties in the nation. This is a result not only of the expertise of the farmers in handling crops and livestock but also of the natural fertility of the soil materials left by glaciers, meltwater, and wind. The fact that dolomite, sandstone, and shale derived locally, as well as igneous and metamorphic rock derived from the north, were ground down and mixed by glaciers and subsequent water and wind accounts for a high natural fertility in the soils. In addition, abundant groundwater in the drift and in the sandstone aquifers permits irrigation of special crops. The soils that we use for agriculture have developed since glacial time by the weathering or breakdown of the minerals contained in the glacial deposits and in the loess. Humus was added in large amounts by native prairies and open woodlands and continues to be contributed by dairy agriculture.

The weathering or disintegration process--acting on the parent materials of till, sand and gravel, and loess--and the incorporation of organic matter produce layers called soil horizons. These horizons form more or less parallel to the ground surface and vary in thickness and characteristics, depending on drainage, position in the landscape, vegetation, and other biotic agents and parent material. Their distribution and character in Dane County are shown in Figures 17 and 18.

For convenience in mapping and describing soils, soil scientists have developed a nomenclature for these horizons, which is shown in Figure 17. The A1 horizon is the upper layer of the mineral soil and it is enriched in organic matter. In poorly drained areas, the A horizon is especially thick (Area 10 in Figure 17). Rather thick A horizons also occur in areas formerly covered by prairie vegetation on well-drained uplands, such as the soils of Areas 2 and 3 in Figure 17). A horizons of intermediate thickness are found in soils developed under savanna or oak-opening vegetation (Areas 6, 8, and 9 in Figure 17).

In areas that were dominantly covered by forest at the time of settlement, a somewhat pale E (or A2) horizon developed just below the A horizon. This paler horizon is lighter in color than the horizons above and below it, and it is a layer that has lost calcium and other mobile elements during the soil-forming process.

Below this, the yellowish or reddish brown B horizon is a layer of accumulation of clay and other weathering products, and also of some organic matter, all of which washed down from above. Thus, this horizon is usually richer in clay-sized particles and colloidal humus than are the upper horizons. Below the B horizon is the C horizon, which consists of oxidized and only slightly weathered geologic ("parent") material, that has not received significant additions of materials from above.

Soils differ in their characteristics from one position in a landscape to another (Figure 18) because of differences in (1) parent material, (2) amount of organic matter incorporated into them under various plant communities, and (3) degree of dryness or wetness of sites. During much of the Holocene, and even up to the present, Dane County has been subject to changeable environmental influences characteristic of this transition zone that lies between prairies of the west that extend across the Great Plains and the forests to the east and north.

Recurrent fires apparently had a great impact on the vegetation and indirectly on the soils. Where the fires, set by lightning and by Indians, were most frequent, tree growth was prevented and prairies flourished. Under these grasslands the dark, fertile prairie soils, represented by Plano and Edmund silt loams (Areas 2 and 3 in Figures 17 and 18) developed in a loess cover over glacial till east of the end moraine, and over bedrock to the west of the moraine. Where fires were less frequent, open oak woodland existed, with prairie grasses between the trees. Under these savanna conditions, the somewhat less dark Basco, Batavia and Meridian soils (Areas 6, 8, and 9 in Figures 17 and 18), among others, formed in loess and loamy covering over glacial drift or bedrock, depending on the site. Where fires were infrequent, deciduous forest predominated and Dodge, Dunbarton, Derinda, and Elkmound (a stony Lithosol) soils (Areas 1, 5, 4, and 7 in Figures 17 and 18) formed in the various geologic materials indicated in Figure 17.

The suburb of Maple Bluff, northeast of Madison, is a site where the forest was protected by Lake Mendota from recurrent fires being fanned by southwest winds. Particularly well-developed deciduous forest soils are found there. Valley bottoms are occupied largely by poorly to somewhat poorly drained dark alluvial soils (Area 10 in Figure 17). In large valleys, such as the lower reaches of Black Earth Creek and the Sugar River, terrace surfaces are occupied by prairie (near Black Earth) and forest (near Belleville) soils. The numerous wetlands in Dane County, including kettles, contain the peats and mucks (Histosols) that developed from plant debris.

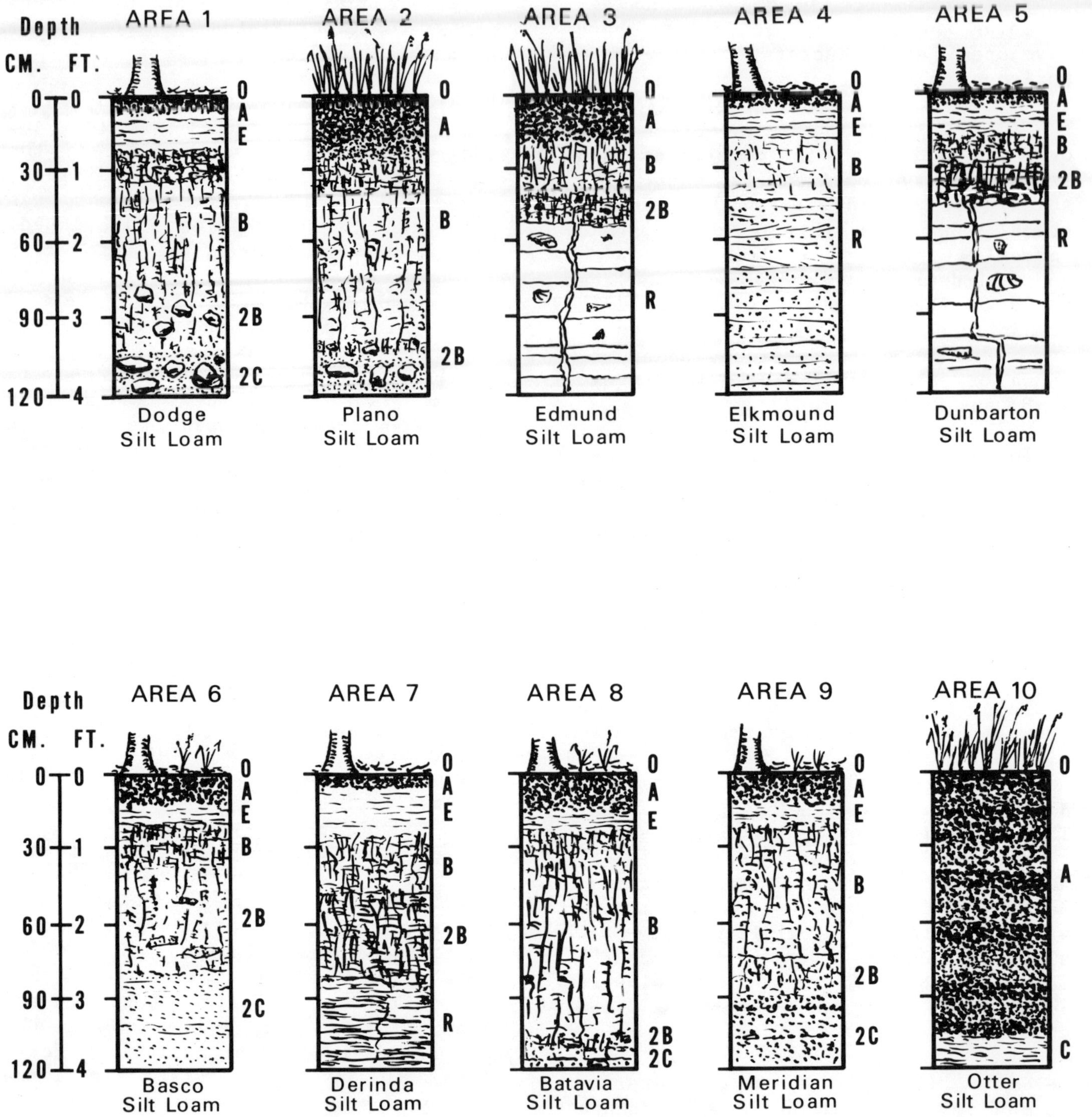

Figure 17: Representative soil profiles from Dane County Area shown at top of each profile corresponds to areas mapped in Figure 18. Diagrams by Francis Hole.

Names of dominant soils developed on various parent materials under deciduous forest or prairie vegetation.

	Loess over Till	Loess over Sand or Gravel	Loess over Dolomite	Loess over Sandstone or Shale
Deciduous Forest	Area 1 Dodge St. Charles McHenry	Area 9 Meridian Granby Dickinson	Area 5 Dunbarton New Glarus Seaton	Area 4 (Sandstone) Elkmound Dunbarton Area 7 (Shale) Derinda Dunbarton
Prairie	Area 2 Plano Ringwood Griswold	Area 8 Batavia Houghton Dresden	Area 3 Edmund Sogn Port Byron	Area 6 Basco Elkmound Gale
Poorly Drained	Otter, Orion, and Troxel soils often formed on stream or lake sediment.			

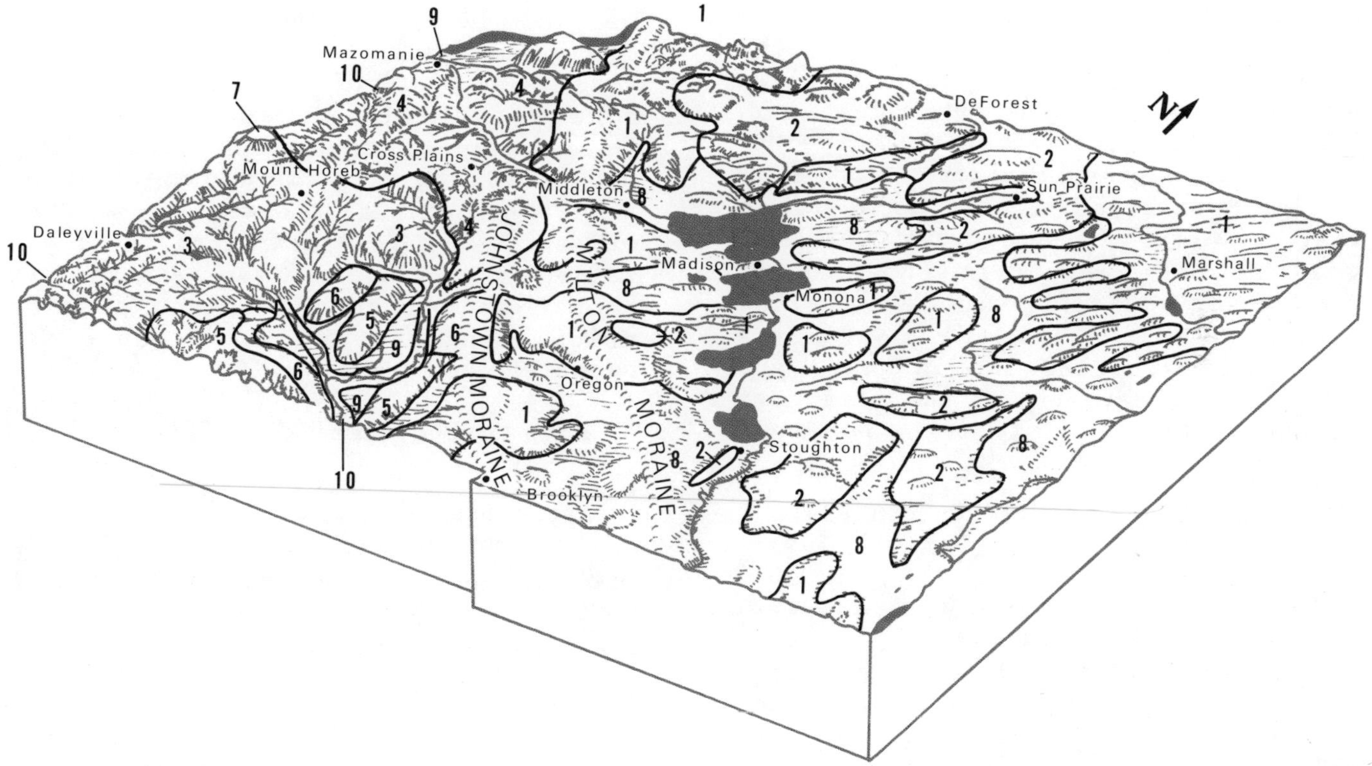

Figure 18: Map of generalized distribution of soil types in Dane County. Figure 17 illustrates the type of soil found in each of these areas. Diagrams by Francis Hole.

This trip is designed to illustrate geologic and landform features of western Dane County. On the trip, you will see various glacial features, such as drumlins, moraines, outwash plains, and lake plains and you will also be able to see two of the primary rock types in the Madison area. The routes of the trips are shown inside the front and back covers. Because many of the features are on private land, I have attempted to designate places where features can be seen from the road. If you do enter private land, especially pits and quarries, remember that permission is necessary. Many quarries are very dangerous and they should not be entered unless for a specific purpose and then only with specific permission from the owner. Since many of the features in Dane County have been described in the text, and the glossary and map explanation provide further insight into the formation of the features, their genesis will not be discussed here. Because many of the exposures are of a transitory nature, the trip will concentrate mostly on views of landscape and geomorphic features. It is possible that after the guide has been written other exposures will show things better and it is also possible that some of the stops recommended here will not be appropriate because of changes of roads, development, or the growth of vegetation. Keep your eyes open and, with the map in hand, you should be able to figure out what many of the features are that are not mentioned specifically in the guide.

To get the most from the trip, I recommend at least 2 people in the car--one to interpret the glacial map and navigate and the other to watch the road! Be very careful if you stop or slow your car along the road. Although I have tried to keep most of the route on lesser-used roads, you should always be alert for other vehicles. Another point to consider is that the season of the year will, to some extent, determine what you can see. The roadlog was laid out in spring, not in late August when corn is higher than your car!

Both trips begin and end at Law Park, along the northwest side of Lake Monona on Highway 151 (John Nolan Drive). Each trip will take a minimum of 4 to 5 hours and I suggest that you bring a lunch and make a day of it. County parks are indicated on the route map and there are other parks in towns along the way. Finally, there is no reason why parts of the trips cannot be done at different times. Mileage between each observation is given along the left margin and I recommend using these when driving the trip. The cumulative mileage given along the right column is only approximate because most people will find diversions or side trips along the way and because car odometers vary.

Have a good time!

Self Guided Field Trip to Western Dane County

(see inside front cover for route map)

0.0

Leave the west parking lot of Law Park along Lake Monona just below the capitol square. The flat land on this side of the square is underlain by till and deposits of glacial Lake Yahara. Head southwest (left on John Nolan Drive).

<u>0.3</u>

0.3

Turn right on North Shore Drive.

<u>0.4</u>

0.7

Continue to the right on Proudfit Street.

<u>0.2</u>

0.9

Continue straight on Regent Street.

<u>1.0</u>

1.9

Turn left (southwest) onto Monroe St. at the corner of Regent and Monroe Streets near the UW Field House. As you rise in elevation to the stop light you leave deposits of glacial Lake Yahara and climb onto the ground moraine. To the right is a drumlin that makes up the core of the University Heights neighborhood.

0.6

2.5

Edgewood High School on the left sits atop a drumlin.

0.4

2.9

You are now back on deposits of glacial Lake Yahara. To the left is Lake Wingra, one of the remnants of the formerly larger glacial Lake Yahara.

0.6

3.5

Continue left on Nakoma Road at its intersection with Odana Road. You are skirting the edge of the ground moraine (to the right) and deposits of glacial Lake Yahara are on the left.

1.2

4.7

Turn left (south) on Midvale Boulevard, which becomes Verona Road, and pass under the west Beltline on Highway 18 and 151 west and south.

0.5

5.2

You are now rising on the up-ice side of the Milton Moraine, the only major recessional moraine formed by the retreating ice. The surface of the moraine has been considerably modified here but if you look carefully, you can still see evidence of closed depressions, called kettle holes.

1.2

6.4

You are now descending from the Milton Moraine onto outwash that extends from here toward the southwest and the town of Verona, where it joins the Sugar River drainage. Note that extensive sand and gravel mining has been done on this outwash surface as we follow it to Highway PB.

0.3

6.7

Intersection with Highway PD. Stay on Highway 18 and 151. To the right is the quarry of Wingra Stone Company. The upper unit being mined is the Galena-Platteville dolomite. Underlying this in part of the quarry is the St. Peter sandstone. Exposures on the left in the next mile show Galena-Platteville dolomite on the ridge top underlain by St. Peter sandstone.

2.6

9.3

Turn left (south) on Highway PB. Here we cross the outwash in the valley and climb onto ground moraine.

1.2

10.5

You are now rising on the up-ice side, or back side, of the Johnstown Terminal Moraine.

0.4

10.9

Turn right (west) on Highway M. Note the kettles on the moraine to the southeast of the intersection. After rising a small distance we will descend the front edge of the end moraine where excellent kettles can be seen.

0.7

11.6

You are now driving parallel to the moraine front. Meltwater flowing off the ice margin carried gravel a short distance from the ice and deposited it on the relatively flat surface you see to the left ahead.

0.3

11.9

To the right and just ahead is a stream valley which drained water flowing from the Milton Moraine at a somewhat later time. Note the gravel pits on the right and left ahead in the outwash.

0.2

12.1

You are now beginning to rise on the end moraine again. Note the large gravel pits on the left which are now water-filled. Continue on M northward through Verona.

0.8

12.9

At the intersection of Highway M with Highway 18 and 151, continue straight ahead on Highway M. You are leaving the end moraine and going onto an outwash surface. Note how flat the surface is in the next mile to the north.

0.8

13.7

Turn left (west) on Cross Country Road. Just before making the turn, you begin climbing again onto the Johnstown Moraine. Note the very kettled surface.

0.6

14.3

Kettle holes are present on both sides of the road here as we begin to descend the front of the end moraine. Note that the crest of the moraine is wooded as it is in many parts of the county. Ahead of us is a small outwash plain and then bedrock hills of the Driftless Area.

0.6

14.9

Turn right (north) on Nine Mound Road. You are now on the outwash and the moraine lies to the right (east). Note the gravel pits that are present here because of the deposits of coarse aggregate. If the exposures are good you can see the difference in size of rocks in the gravel in pits away from the moraine compared to those near the moraine without leaving your car as you drive along.

0.7

15.6

Turn left (west) on Highway PD. A large sand and gravel operation is present south (left) of the road. You can glance back and see that the moraine is a fairly impressive ridge north and east of here.

0.3

15.9

Turn right (north) on Shady Oak Lane. You remain on the outwash. The moraine is ahead.

0.4

16.3

As the road turns to the left it begins to parallel the moraine front. After the road turns to the right, the south-flowing drainageway that you see to the left steepens and gets deeper as you continue toward the north. This drainageway was the source of some of the water that deposited the sand and gravel you just saw.

0.7

17.0

Meltwater cut through the moraine to your right at this location and water flowed through this channel as the glacier margin retreated from the Johnstown Moraine. Richardson Cave, a small cave in the Platteville-Galena dolomite is present a few hundred yards up this drainageway. Continuing on Shady Oak Lane, you now climb onto the Johnstown Moraine.

0.5

17.5

Turn left (west) on Midtown Road. Here, at the crest of the moraine, you leave the moraine and head west into the Driftless Area. The surface here is underlain by several feet of windblown silt (loess) over bedrock.

0.2

17.7

Bear to the left, remaining on Midtown Road.

0.5

18.2

Turn right on Timber Lane. St. Peter sandstone is exposed in the roadside ditches in this area and at several localities within the next mile. Note the lack of erratic boulders.

1.9

20.1

You are still in the Driftless Area. Glance to the right and you will see the wooded crest of the moraine, approximately one quarter mile to the east.

0.6

20.7

Intersection with County S (Mineral Point Road). Remain on Timber Lane. This area is shown in Figure 14 in the text. You are now again along the moraine front but the moraine is smaller here than it was to the south. Note the sharp crest that can be observed behind the farm buildings on the southeast side of the intersection and the wooded moraine crest just east of the road to the north of the intersection. Underlying the road itself is sand and gravel; and to the west are two fairly large kettles that are now partially filled with lake deposits. Water from this area drained northward along the

edge of the ice to Wilke Gorge where it drained into Black Earth Creek. Continue north on Timber Lane. The edge of the moraine remains to the right as you cross the outwash surface and climb onto bedrock of the driftless area.
1.5

22.2

Timber Lane bears left and becomes Cleveland Road.
0.2

22.4

Intersection of Cleveland Road and Old Sauk Pass Road. To the northwest of this intersection is an area that will become one of the sites of the Ice Age National Scientific Reserve. A map of the site location is given in the text (Figure 14) and the history of Wilke Gorge is also discussed. The area is discussed in more detail by Black . Retrace your route east and south along Cleveland Road and Timber Lane to Highway S (Mineral Point Road).
1.7

24.1

Turn right (west) on Mineral Point Road.
0.4

24.5

This road cut is in Galena-Platteville dolomite. Note the appearance of the dolomite in outcrop and contrast it with the St. Peter sandstone that will be seen a short distance further west. From here west you will be in the Driftless Area for quite some time.
1.5

26.0

On the right is an outcrop of St. Peter sandstone, the stratigraphic unit below the Platteville-Galena dolomite (see text Figure 3). Note that the surface of this sandstone is smooth, on the lower part very light colored, and a great deal of loose sand is produced as the soft sandstone weathers. The sandstone often weathers into peculiar shapes because of differential cementation of the sand grains.
0.9

26.9

Turn left (south) on County P in Pine Bluff.
1.2

28.1

Another exposure of St. Peter sandstone. The St. Peter sandstone is mined for industrial sand in the quarry to the left.
1.0

29.1

Bear right on Highway S at the intersection of Highway P and S. On the left behind, note the stream meanders. You are now rising onto a nearly flat upland surface underlain by Galena-Platteville dolomite. To the right, or north, is the valley of Black Earth Creek and to the left (south) are the gentle slopes of the Sugar River drainage.
3.4

32.5

Turn left (south) on Highway 78.
0.8

33.3

Turn right (west) on Highway 151 and 18 and continue west through the town of Mt. Horeb. You are now on the divide between drainage to the north to the Wisconsin River and to the south to the Sugar or Pecatonica Rivers. From here westward to the Mississippi is Military Ridge. This ridge, underlain primarily by Galena-Platteville dolomite, stands above the surrounding countryside and was the location of an early military road from Green Bay and Milwaukee westward to Prairie du Chien. It was also the site of the earliest railroad across southern Wisconsin because, being on the divide, only a few very small bridges across stream valleys had to be built. A discussion of the geology of this area is given in the text and a north-south cross section is given in Figure 4. We will reserve a discussion of the overall geology for a stop at the top of West Blue Mounds.
5.6

38.9

At the town of Blue Mounds, follow the signs to Blue Mounds State Park, turning right (north) off Highway 151 and 18. The town itself is underlain by Galena-Platteville dolomite. Cave of the Mounds, a commercial cave just east of town, is developed in this geologic unit.

0.5

39.4

As the road bears to the left note that the slope is very gentle. We have reached the top of the Galena-Platteville dolomite and the gentle slope represents the base of the overlying Maquoketa Formation which consists of shale and dolomite. Just ahead on the north side of the road is a small group of trees that are in a sinkhole. This circular depression looks like a kettle but it is not glacial in origin. It developed as water percolated through cracks in the underlying dolomite, dissolving the dolomite and enlarging the cracks by solution and perhaps collapse.

0.3

39.7

Continue straight and rise again as you enter the park. Note the scatter of large boulders of the very resistant Niagaran dolomite. This dolomite is very resistant to erosion and is responsible for Blue Mounds standing as a hill above the surrounding countryside. It is resistant because after the dolomite formed, silica, much of it in the form of chert, replaced some of the calcium and magnesium carbonate. In the head waters of streams flowing to the north from the escarpment there is a large amount of this chert in the valley bottoms.

0.6

40.3

When you reach the top of west Blue Mound, climb one of the towers, preferably the eastern one. As you scan the countryside from the top of one of the towers examine Figure 3 in the text. Note that the Wisconsin River valley is to the north and relatively steep, short tributaries drain this side of the escarpment. To the south, much more gentle streams flow into the Pecatonica River drainage and eventually to the Rock River in Illinois and then to the Mississippi. Because of the steep slopes, the north side of the escarpment is forested considerably more than the south side. The streams flowing to the deeply entrenched Wisconsin River cut through a number of rock types in a relatively short distance. To the south, stream gradients are much more gentle and because the rocks are dipping, or tilting toward the south, the streams flow on one rock type for a fairly long period of time. East Blue Mound, the hill directly east of here, is lower and is flat topped. On this mound, the Niagaran dolomite has been eroded away and Maquoketa shale, with some interbedded dolomite, caps this mound. When leaving here you will proceed northward down the relatively steep tributaries into the Wisconsin River drainage. Retrace your route through the town of Blue Mounds.

1.9

42.2

Turn left (east) on Highway 151 and 18.

1.7

43.9

Turn left on Highway JG. You are now leaving Military Ridge or the top of the escarpment and descending the head waters section of a tributary to the Wisconsin River. Note the steep wooded slopes in the headwaters area. The stream valleys here are choked with chert derived from the Niagaran dolomite and the older Galena-Platteville dolomite.

2.2

46.1

Here you enter the valley of Bohn Creek. Note that the stream valley is beginning to widen and a floodplain is developing. In places the stream meanders or winds back and forth on the floodplain. The width of the floodplain and the size of meanders increase in the downstream direction.

0.6

46.7

Turn left on Bohn Road and continue down valley. The bedrock in the lower slopes of the valley here is St. Peter sandstone. The steep upper slopes are underlain by Galena-Platteville dolomite.

0.9

47.6

Along Elver's Creek in this location is the site that was studied by Davis .

Here gravel composed of chert from the dolomite, presumably deposited during glacial time, underlies the whole valley. Above the gravel is silt and peat containing spruce wood dated at 10,480 years. Overlying the peat is more silt deposited during Holocene time. This site is discussed in more detail by Davis. The lower slopes of the valley walls are made up of Prairie du Chien dolomite and the St. Peter sandstone forms the fairly gentle slopes about half way up the valley walls. <u>Do not enter private land.</u>

<u>0.4</u>

48.0

Turn left on Highway J. Note the meanders at various places along the creek as you pass down the valley. In places, on cut banks, you can see the post-settlement alluvium that is discussed in the text.

<u>1.1</u>

49.1

Turn left on Highway J at the hamlet of Elvers.

<u>0.8</u>

49.9

Turn right on Highway F.

<u>0.8</u>

50.7

Cross Ryan Creek. Note that the floodplain in this area has been artificially drained by the straight channel. Many of the floodplains at this elevation were very wet before drainage--partly because high outwash filling in Black Earth Creek Valley and the valley of the Wisconsin River dammed all of these tributaries, creating temporary lakes about 12 to 14 thousand years ago. Since that time, the streams in these valleys have been downcutting slightly through the lake sediments and accumulated alluvium.

<u>1.8</u>

52.5

At the intersection of Highways FF and F continue straight (north) on Highway FF. The bedrock at the base of the valley walls is Cambrian sandstone overlain by Prairie du Chien dolomite. Refer back to text Figure 3 and follow our progress down through the bedrock units.

<u>1.6</u>

54.1

You now enter Iowa County for a few miles before returning to Dane County. Road becomes Highway KK. Continue straight on Highway KK.

<u>0.8</u>

54.9

Intersection with Highway K. Turn right (north) on Highway K. Note that the valley here is quite wide. This is because you are approaching the valley of Black Earth Creek and the Wisconsin River and you are driving over fairly thick backfilled sediments deposited in a lake as the main valleys were filled with outwash derived from the glacier to the east.

<u>2.0</u>

56.9

You can now view the very wide valley of the Wisconsin River ahead. Black Earth Creek joins the Wisconsin River some distance to the west. The surface you are on now is underlain by outwash deposited 14 to 15 thousand years ago when the ice margin was at the Johnstown Moraine. At that time, outwash was carried down Black Earth Creek and a number of tributary valleys as well as the main Wisconsin River valley toward the Mississippi. Because this area was a long distance from the ice margin, the outwash here is sand as opposed to the coarser gravel that we saw adjacent to the moraine. The small hills on the surface are sand dunes that were formed before vegetation was firmly established on the outwash surface. Probably during drier times of the Holocene the sand dunes were also active. Many bare sand areas or blowouts existed on this surface until recently. As you cross the outwash surface you will note a number of places where pine trees have been planted to break the wind and stabilize the surface.

<u>0.4</u>

57.3

Turn right (east) on Highway 14.

<u>0.2</u>

57.5

Turn left (north) on Morrill Road. Note that you descend into the channel of Black Earth Creek from a level some 15 to 20 feet higher than the creek itself. This amount of channel cutting has taken place since the outwash was deposited.

0.2

57.7

Cross Black Earth Creek.

0.3

58.0

Note that you are again rising onto a higher outwash surface.

0.3

58.3

Rise onto upper terrace.

0.4

58.7

Turn right (east) on Hudson Road.

0.8

59.5

Stop at the corner of Beckman Road and look right into the valley of Black Earth Creek. If you look carefully, you will see that there are actually two levels of outwash above the stream bottom. One of these surfaces is the level that you are now on and you can see the matching outwash surface across the valley. An intermediate level corresponds to the field below you and the surface across the valley and somewhat downstream where you can see a house and several barns. These levels are called terraces. They form because the stream carrying outwash cannot carry all of the sediment out of the valley. Thus during glacial time the outwash surface builds up fairly rapidly, perhaps at times several feet per year. After the ice has retreated, the size of the stream is reduced, and instead of being a whole series of shallow channels flowing across the outwash surface, water is confined to a single channel. When this occurs the stream starts to downcut or erode through the outwash. Any pauses in this downcutting create lower terraces. It is postulated that the high outwash terrace that you are on now was formed during the glacial maximum when ice was at the Johnstown Moraine. Following that, the stream downcut rapidly, then paused at the intermediate terrace, probably as meltwater again flowed down the Wisconsin River valley. After ice retreat from the Green Bay lowland, subsequent stream cutting has taken place to the elevation of the present stream channel. Turn left (north) on Beckman Road.

0.2

59.7

Note as you drive along that the soils are very sandy and that there is very little loess covering the terrace surface. If loess was deposited here, much of it was probably blown off during dry periods during the Holocene. Some loess is present on the upper terrace but there is very little present on the intermediate terrace, because this terrace was cut after loess deposition ceased.

0.6

60.3

You are now descending from the upper terrace onto the intermediate terrace. The difference in elevation is approximately 50 feet. Note the sand on the right that underlies the terrace.

0.4

60.7

Turn right (east) on Amenda Road. A small sand dune is located on the right and you cross another a tenth of a mile ahead. Although some of this area looks wet it is because of a high water table and the soils become very droughty if the water table is lowered.

0.8

61.5

Small sand pit to left at right angle turn. To the left is the flood plain of the Wisconsin River which is between 10 and 20 feet lower than the intermediate terrace that you are on now.

1.1

62.6

Turn right (south) on Highway Y. You are still on the intermediate terrace.

0.3

62.9

You are climbing onto the upper level terrace. Note the exposure of sand to the left. This is typical of the interior of the terraces.

0.7

63.6

You are descending from the upper level terrace into the valley of Black Earth Creek.

0.4

64.0

Cross Black Earth Creek.

0.1

64.1

Turn left (east) on East Hudson Road.

0.2

64.3

Cross Black Earth Creek. You now rise onto the intermediate terrace.

1.0

65.3

Turn left (northeast) on Highway 19 and 78. You are now climbing onto the upper-level terrace but it is difficult to see ahead because of the bedrock hills.

0.8

66.1

Turn right (east) on Highway 19. The gravel in the pits on the left is somewhat coarser than the sand that we saw west of here. This gravel was derived from Halfway Prairie Creek, which you will follow as you proceed eastward on Highway 19.

1.0

67.1

Intersection with Highway F. Continue eastward on Highway 19. This whole valley is underlain by outwash. Many of the soils were poorly drained before artificial channels were cut. Cambrian sandstone is exposed ahead and to the left.

2.2

69.3

The outermost extent of glaciation in this area is very difficult to identify. Presumably the ice crossed the valley at about this location but it left no terminal moraine as it did to the south and to the north. On the uplands you can trace the distribution of erratic boulders (those carried into the area by glaciation) but in many places the boundary is indefinite. The surrounding hills are made up of Cambrian sandstone and Prairie du Chien dolomite.

1.0

70.3

Note the often wet kettle holes in this area of pitted outwash. You are west of the glacial border.

0.8

71.1

Several gravel pits are in this surface just west of Marxville. You are now within the glacial border so that the gravel is relativly coarse. Continue east on Highway 19.

1.5

72.6

You are passing Indian Lake, a kettle hole, on the right. Note as you proceed that the valley narrows very rapidly.

2.2

74.8

Stop about 0.2 miles beyond Whipporwill Road. Behind you, in the narrowest part of the valley, is part of the old preglacial divide which separated drainage to the Wisconsin River and drainage to the Yahara River system (Fig. 2). Although glacial meltwater has modified the divide and cut the valley, the location remains a high point. To the east one can see a depression surrounded by hills and containing Lake Brandenburg. In the depression are some kettle holes and kames which were formed along the ice margin as ice retreated from the Milton Moraine. On the uplands north and south of here remnants of the Milton Moraine can be seen. This large semicircular depression that you are now in, surrounded by steep bedrock walls, was probably greatly enlarged by subglacial drainage. Drill records indicate that there are two hundred feet of glacial deposits in the valley bottom. The size and depth of the valley here are impressive because the valley was much smaller in preglacial time. Because subglacial meltwater flows under pressure, carrying boulders and gravel, it can erode bedrock and other underlying materials and also flow up hill much as water does in a pipe. Thus, water could have drained out of this deep basin, across the narrow divide in which you are standing, and down the valley which you have just driven through. As we progress eastward from here we will see numerous indications of subglacial drainage. In some cases the indications will be large valleys which presently contain very small streams. Another indication will be deposits of ice-contact stratified drift, or kames and kettles, that indicate that water-transported material was deposited partially on ice. Thus one could envision a large stream flowing beneath the ice and actually cutting into the bedrock in some areas and flowing up in the ice and depositing stratified material such as sand and gravel on perhaps several hundred feet of ice in other areas. As the glacier retreated and the underlying ice melted away the sand and gravel collapsed into typical kame and kettle topography. Continue east on Highway 19.

0.8

75.6

A kame composed of sand and gravel in valley bottom on left.

0.2

75.8

Turn right (southeast) on Highway 12 and 19, remaining in subglacial channel.

0.3

76.1

Kame on left side of road.

0.5

76.6

Turn left (northeast) on Highway P. Before turning, look to the right and notice the kame and kettle topography with small gravel pits. The surface you cross in the next mile is entirely underlain by ice-contact stratified drift as you remain in the subglacial channel.

1.0

77.6

Turn right on Hyer Road (second right). We are now descending into part of the subglacial drainage system that was a channel. In the low areas, note the flat surface with poorly drained soils that probably represents former shallow lakes in this basin after ice retreated. By examining the glacial geology map you can see that a similar valley is adjacent to us to the south and it is in this valley that Waunakee Marsh has developed. Across the valley you can see gravel pits in ice-contact stratified drift which in many places borders the edge of the valley indicating that some stratified materials were deposited on and against ice here. (If ice were not present in the valley bottom when the gravel was deposited, the gravel would have been deposited on the valley floor, not against the hillside.)

1.2

78.8

Rise out of the valley and note the gravel pits on the right in ice-contact stratified drift.

0.4

79.2

Turn right (south) on Kuehn Road. You are now on outwash that has several shallow kettle holes in it and so it is mapped as pitted outwash. This outwash seems to have been derived from the valley to the north as the ice was receding.

0.5

79.7

Turn left (east) on Highway 19. To the south lies Waunakee Marsh, another branch of the subglacial drainage system. Note the kames and numerous small former gravel pits along this road at the edge of the marsh.

0.8

80.5

The lowest slopes of this hill are Prairie du Chien dolomite.

1.1

81.6

To the north and ahead of you is outwash deposited as the ice receded from this area.

0.2

81.8

Rise onto ground moraine.

1.1

82.9

Turn left (north) on Highway 113.

0.2

83.1

Cross Six Mile Creek. You are now on the outwash surface again. Contrast this surface with the hilly ground moraine surrounding you. One of the valleys that carried subglacial meltwater lies to the right, ahead.

0.9

84.0

Rise onto ground moraine. Several drumlins are present ahead of you.

0.5

84.5

Descend into a valley that now contains no stream but that was cut by the subglacial drainage system.

0.5

85.0

Turn right (east) on Cuba Valley Road. We climb onto ground moraine almost immediately and cross a drumlin at the crest of the hill.

0.5

85.5

You are now descending into the easternmost valley that carried subglacial drainage.

0.5

86.0

You rise again out of the valley and continue on ground moraine.

0.3

86.3

Cross Schumacher Road. The subglacial valley continues on the left.

0.7

87.0

Turn left (north) on Patton Road and recross the valley again before rising onto ground moraine.

0.6

87.6

As you begin descending this hill, probably underlain by bedrock, look to the northeast and notice the amount of wooded humocky topography. This is an area of ice-contact stratified drift which was deposited in the subglacial tunnel.

0.9

88.5

Turn right (east) on Norway Grove School Road. You are now in the middle of the ice-contact stratified drift and as you proceed along the road you will note kames and kettles and numerous gravel pits. Aggregate for asphalt and for ready-mix concrete is being produced in these operations. Without leaving the car note the characteristics of the gravel exposed to the north of the road. Although the pit face changes with time, normally the gravel here is collapsed and very variable in composition. It does not show the continuous and uniform stratification that we have seen in outwash deposits.

1.0

89.5

Turn right (south on Highway I).

0.5

90.0

Rise onto ground moraine leaving the ice-contact stratified drift deposits. You will remain on ground moraine, which consists of rolling topography with relatively thin till over bedrock, until south of the town of Waunakee. Remain on Highway I to the intersection of Highway 113.

3.6

93.6

Turn right (west) on Highway 113 and 19.

1.0

94.6

Look south across the Yahara Lowland.

1.6

96.2

Turn left (south) on Highway Q after passing through the town of Waunakee. The area to the south is all ground moraine and you begin to see some of the lineation of the land surface due to ice flow from northeast to southwest.

3.0

99.2

Turn right (west) on Highway K. Before turning note the gravel pits to the left in outwash. A drumlin is present to the northeast and northwest of this road intersection.

0.2

99.4

Climb onto drumlin.

0.8

100.2

Turn left (south) on Highway K at the intersection of Pheasant Branch Road.

0.5

100.7

Turn right on Highway K. Gravel pits to the left are in outwash.

1.4

102.1

Remain on Highway K across Highway 12. You are still on ground moraine.

1.2

103.3

Village of Ashton. Turn left (south) on Church Road.

1.0

104.3

Turn right (west) on Schneider Road.

0.5

104.8

Turn left (south) on Capital View Road. You have just climbed from outwash onto ground moraine.

0.5

105.3

Cross small area of ice-contact stratified drift deposits and note the kettles, some of which are water filled. The flat surface to the left is underlain by the deposits of glacial Lake Middleton.

0.5

105.8

After rising onto ground moraine turn left (east) on Airport Road.

0.3

106.1

You have descended onto a very flat surface underlain with lake sediment. The lake that was present was called glacial Lake Middleton and it drained westward down Black Earth Creek. Although the deposits are very thick (over 100 feet) the lake was shortlived and it existed only for the period of time that glacier ice blocked the central and eastern parts of the Lake Mendota basin. As soon as deglaciation had occurred in the University area and eastward to the Capitol, the lake level dropped and the lake sediments here were exposed and vegetated. Because of the fairly large amount of silt and clay in the sediment, and because of the flat surface, the soils are poorly drained and many are organic. As you cross the surface you will note that drainage ditches have been placed to make agriculture possible. If you take this trip in the Spring of the year you can see a band of light colored soils around the edge of the lake deposits. These light colored soils are sandy and represent the beach deposits of the old lake. You will remain on deposits of glacial Lake Middleton until we descend into the valley of Pheasant Branch Creek.

1.9

108.0

Intersection with Highway 12. Continue straight on Highway M.

0.8

108.8

You are now descending into the valley of Pheasant Branch Creek and onto lake sediments deposited later than those in glacial Lake Middleton. This relatively flat surface has been extensively modified by drainage and construction. Before modification, however, much of the surface was at an elevation of about 860 feet and these deposits were left by glacial Lake Yahara, a lake which extended from here to Stoughton at an elevation of about 860 feet, or about 12 feet higher than the present level of Lake Mendota (see text for further discussion).

0.5

109.3

Turn right on Highway Q. You continue on deposits of glacial Lake Yahara. Note, as you drive along this section of road, the flat surface that has been somewhat modified by filling but still represents shallow water deposits of glacial Lake Yahara.

0.8

110.1

Rise onto ice-contact stratified drift.

0.2

110.3

Turn left (southeast) on University Avenue.

2.0

112.3

Descend onto outwash in the Yahara Lowland. You will remain on outwash as you continue on University Avenue,

Campus Drive, and Johnson Street respectively for about 3 miles.
<u>2.9</u>

115.2

Near the intersection with Lake Street you descend slightly onto deposits of glacial Lake Yahara. Continue on Johnson Street.
<u>0.3</u>

115.5

Turn right (southeast) on Bassett Street.
<u>0.2</u>

115.7

Cross West Washington Avenue.
<u>0.3</u>

116.0

After joining West Wilson Street, turn right (southeast) on Broom Street.
<u>0.1</u>

116.1

Turn left (northwest) onto John Nolan Drive.
<u>0.1</u>

116.2

End of trip at west parking lot of Law Park, along shore of Lake Monona.

Self Guided Field Trip to Eastern Dane County

(see inside back cover for route map)

0.0

Begin at Law Park on the northwest shore of Lake Monona on John Nolan Drive in Madison. To the north is the Capitol, which is atop a drumlin. Head west on John Nolan Drive along fill that has been placed over lake sediments. Note that the surface of the road is not really flat. This is because lake sediment and organic material beneath the fill are unstable and they move differentially because of the load above.

0.9

0.9

As you leave the causeway over Lake Monona you cross onto lake sediment of glacial Lake Yahara. Much of this area has been filled and most of it was marsh at one time. The area beneath the coliseum (ahead) has all been filled.

1.3

2.2

Turn east on the Beltline (Highway 12 and 18). As you enter the Beltline on the ramp, the low area to the right was also once part of glacial Lake Yahara. This provided a connection between Lake Monona and Lake Waubesa. Nob Hill (to the right) is a drumlin.

1.5

3.7

Intersection of Highway 12 and 18 and Bridge Road. There are two potential stops to examine old shorelines that indicate more extensive lakes. If you drive north on Bridge Road to the Yahara River, a bike trail runs north from Bridge Road along the lake. If you walk the bike trail, about 200 yards north and then go off the trail to the right, you'll find a concentration of boulders, which is a record of higher lake level. The same kind of feature is more accessible and better developed just south of the Beltline in back of Lum's restaurant. Turn off the Beltline, drive into the T intersection, turn left and stop at the edge of the marsh. Walk south along the edge of the marsh and you will find a concentration of boulders that have been washed out of the till by wave action. This shoreline is relatively low and does not represent the highest stand of glacial Lake Yahara. It is not clear when this shoreline formed and it may have been in the relatively recent past before organic materials were well established. Be sure to readjust your mileage after this stop.

0.4

4.1

From the intersection of Bridge Road, continue eastward on the Beltline. As you cross the Yahara River look south and observe the large flat plain across the river. This is all lake sediment, some of which is overlain by organic materials. The area is flat because of the former presence of the lake.

1.3

5.4

Turn right (south) on Highway 51 toward McFarland.

0.6

6.0

Turn right on Terminal Road and go south. You are still on the flat lake plain and remain on it for about 2 miles until the storage tanks have been passed and we rise onto the upland. The hills to the left in back of the storage tanks are drumlins and drumlins are present in the landscape in front of us.

1.4

7.4

Turn left on Taylor Road, just across the bridge. After going under Highway 51, on the left and right drumlins are present. Small roadcuts show the type of material that is present in the drumlins. Both of these drumlins have some sand or sand and gravel beneath, but they are made up primarily of till. They are elongated in the northeast-southwest direction, indicating the direction of ice flow.

0.4

7.8

Turn left on Paulson Road. Small exposures are present along the dirt road to the left. The material in the drumlins is till that is typical of the Dane County area. Note that it is made up of mixed grain sizes such as sand, silt, and clay with some boulders present. Adjust your mileage at the corner of Paulson and Taylor Roads and continue east on Taylor Rd.

<u>0.5</u>

8.3

Continue right (southwest) on Taylor Road and right (west) again on Farwell Street to return to Highway 51. About 0.4 mile after turning onto Farwell Street you drop off the till upland onto the deposits of glacial Lake Yahara.

<u>0.5</u>

8.8

Turn left (south) on Highway 51. All of the hills in this area are drumlins, some of which are completely surrounded by lake sediment.

<u>0.5</u>

9.3

Rise onto ground moraine. Dolomite is relatively close to the surface in this area.

<u>0.9</u>

10.2

Look left at lake plain of glacial Lake Yahara.

<u>0.4</u>

10.6

Cross sediments of glacial Lake Yahara.

<u>0.3</u>

10.9

Rise onto ground moraine. Note drumlins on both sides of the road. Some of the roadcuts between here and Stoughton show dolomite is present. From the higher points of this upland one can see the Milton Moraine to the south. It is the wooded ridge on the horizon. Lake Kegonsa is to the left or east.

<u>1.5</u>

12.4

Turn right (west) on Highway B. In this area bedrock is very shallow and most of the hills have a bedrock core. This area is mapped as ground moraine on the glacial geology map and you can sense the elongation of hills due to ice flow from northeast to southwest. From the hilltops along Highway B you can look to the south (left) and see the Milton Moraine, built during ice recession, running more or less parallel to the road about a mile to the south (on the horizon).

<u>4.6</u>

17.0

Turn left (south) on Highway MM. As you approach the moraine, note the wooded slopes on either side of the road. The road here passes through a fairly low area in the moraine, but the ridge is higher to the right and left.

<u>1.5</u>

18.5

You are now descending from the moraine front onto the outwash plain in the vicinity of Oregon. Note the old gravel pits on the left. Gravel pits are typically located just in front of end moraines because this is where the coarsest aggregate material is available.

<u>0.6</u>

19.1

Just beyond intersection with County M, stop the car and look back along the moraine front. Note the row of trees that delineates the most hummocky part of the moraine. The surface you are on is called pitted outwash. It is material deposited by streams running away from the glacier while it was at the moraine. Here it was deposited around ice blocks that had been left by the retreating ice. Subsequent melting has produced the depressions. Unlike the moraine, which also has kettles, pitted outwash retains much of its original flat surface. Continue south on Main St. (Highway MM) into Oregon.

<u>1.4</u>

20.5

Continue south through the town of Oregon but instead of bearing left on Janesville Rd. (Highway MM) continue south on South Main Street.

<u>0.8</u>

21.3

Turn right (west) on Lincoln Road. You are now crossing ground moraine where bedrock is relatively close to the surface. To the left and ahead lies the Johnstown Moraine (not visible).

<u>1.3</u>

22.6

At the hill top, look to the right and slightly behind--Milton Moraine is ridge on horizon. The area you drove through north of Oregon is just to the right of yellow buildings with red roofs about 3 miles away. The Johnstown Moraine is several miles ahead (beneath the radio tower to the northwest).

0.7

23.3

You are descending into an area of pitted outwash. Note the kettle that the road crosses. To the right are a number of kettles including one that contains Lake Harriett (barely visible). The ridge to the south and ahead is the Johnstown Moraine.

1.0

24.3

Enter Johnstown Moraine.

0.1

24.4

Turn left (south) on Tipperary Road. You are now passing through a low point in the end moraine. Note the kettles on both sides and the litter of boulders on the till surface. Small deposits of ice-contact stratified drift are also present in the moraine.

0.6

25.0

You are now at the outer edge of the moraine on a somewhat pitted outwash surface. Note the presence of kettles west of the road. There are two fairly large gravel pits about one-half mile west of here on this outwash surface.

0.5

25.5

Here at the crest of the hill is an exposure of St. Peter sandstone. This sandstone is Ordovician age and was deposited in a shallow sea almost 500,000 years ago. Note the stratification or layering that is typical of sedimentary rocks like this one. Closer to the end of your trip you will see dolomite, the other major rock type exposed in Dane County. The area that you are in now was glaciated sometime previous to the formation of the Johnstown Moraine. This is actually the northernmost extent of the Brooklyn Moraine which passes from here more or less parallel to the Johnstown Moraine, south to the town of Brooklyn. There is actually very little moraine form but areas of kettles such as to the south of the hill you are on are fairly common. Note that some erratics are present on the surface.

0.2

25.7

Continue south on Tipperary Road.

0.2

25.9

Turn left (east) on Highway A. You are now crossing outwash derived from the Johnstown Moraine, which lies north and directly ahead of us. After skirting a bedrock knob with thin till and loess on it, you will rise on a bedrock surface and then onto the moraine.

1.2

27.1

You are again rising to the crest of the Johnstown Moraine. Note kettles and boulders. You can look back (westward) to the low outwash surface of Story Creek. Outwash and meltwater were carried down Story Creek and into the Sugar River drainage when ice was at the Johnstown Moraine. Presumably, as soon as ice started to retreat from the Johnstown Moraine, drainage was diverted toward the southeast and flowed along the surface where the town of Oregon is located. The moraine is particularly high here because it is on a bedrock hill. As we descend the up-ice side of the Johnstown Moraine, note that the topography is somewhat kettled, but that the slope is not nearly as steep as the moraine front. This is typically the case with moraines in the midwest.

1.0

28.1

At the valley bottom we are entering a pitted outwash plain. Note the shallow kettles north and south of the road on the otherwise flat surface. Continue east on Highway A.

1.0

29.1

Turn right (south) on Union Road.

0.2

29.3

Turn left (east) on Rome Corners Road. The Johnstown Moraine is to your right and you are entering an area of ice-contact stratified drift and pitted outwash.

1.2

30.5

Cross Highway MM. The road continues along the edge of the Johnstown Moraine beyond its intersection with Highway MM to Highway 14. The area to the left (north) of the road is ice-contact stratified drift (kames and kettles) made up of sand and gravel and trending (see the Dane County glacial geology map) north-south instead of northwest-southeast as the moraine does. This ice-contact stratified drift was probably deposited in a tunnel or set of tunnels in the glacier ice by streams draining water out of the glacier. Compare this to the area near Utica which you will be seeing later on this trip and which is of a similar origin. Note that the Milton moraine not only has kettle topography, but is much more wooded than here and has a number of boulders on the surface which are lacking here. The ice-contact stratified drift is somewhat kettled but generally lacks the boulders because it is water deposited material.

0.8

31.3

Turn left (north) on Highway 14. A small area of organic soils, probably overlying lake sediment, lies to the east. You then climb onto ice-contact stratified drift again. Note again the fairly large number of kettles that are present.

2.7

34.0

Turn right (northeast) on Highway 138. To the right is a small patch of morainal topography and to the left is the gently sloping surface of outwash deposited by streams flowing away from the receding ice at the Milton Moraine. The Milton Moraine is on the horizon to the north. All of the flat-lying area has a cover of organic material and fine-grained silt and clay that has been washed in during the last 10,000 years.

2.0

36.0

Turn left (north) on Sunrise Road. This area is ground moraine.

0.7

36.7

Note the 50-foot-deep channel on the right side of the road. This channel drained water from the outwash surface around Oregon out to the east through the Badfish Creek. The cutting of the channel took place after the deposition of the outwash although some of the water that cut the channel may have been derived from melting ice. A short lived lake may have existed in the area to the west of the channel.

0.2

36.9

Turn right (east) on Rutland-Dunn Town Line Road. As we drive over till and outwash surface, note the tree-covered Milton Moraine to the left and the scattered kettle holes on the pitted outwash.

1.8

38.7

Intersection with Hawkinson Road. The surface you are on is what is called pitted outwash. It is material deposited by streams running away from the glacier while it was at the moraine. Here it was deposited around ice blocks that had been left by the retreating ice. Subsequent melting has produced the depressions like Indian Lake which you can see to the south. Unlike the moraine, which also has kettles, pitted outwash retains much of its original flat surface. Look to the right, ahead and behind, and note the remnants of the former flat surface. Continue east on Town Line Road.

0.1

38.8

Stop the car at the top of the first rise and look back along the moraine front. Note the row of trees that delineates the most hummocky part of the moraine.

0.2

39.0

This gravel pit is the pitted outwash along the moraine front. Many gravel

pits are located in this topographic position.
0.2

39.2

Intersection with Lake Kegonsa Road. Continue left on Town Line Road. Enter the moraine again where it is a very low feature. It is low here because thicker glacier ice was here and much of the debris coming out of the glacier ice was carried out onto the outwash plain to the south. Later, when glacier ice melted, much of the moraine was kettled and left at a fairly low elevation.
0.3

39.5

After following the road to the north, continue following Town Line Road to the east around a fairly large kettle hole that is now partially filled with peat. During some seasons of the year, there is sufficient water to cover the peat surface. Continue eastward after leaving the kettle hole. You are still actually in the Milton Moraine but morainal topography is not very obvious here.
1.3

40.8

Turn right (south) on Highway 51. Look ahead and left, as you approach the stop sign, at the Yahara lowland in the distance and the drumlinized upland to the east. Continue through Stoughton.
2.0

42.8

Join Highway A. Continue on Highway 51.
1.1

43.9

Turn right (south) on County Highway A on the east side of Stoughton. Most of this surface is underlain by pitted outwash.
1.6

45.5

Turn right (south) on Collins Road. This area is also pitted outwash and there are two large gravel pits just out of sight to the east.
0.8

46.3

At the 90-degree turn in the road, stop and look to the west. This flat surface is the southernmost extent of glacial Lake Yahara. Note that its appearance is similar to the areas to the north that are also underlain by lake sediment. You are now in the drainageway of the former lake. A detailed map of this area is shown in Figure 16 in the text. The present valley of the Yahara River is about a mile and a half to the west. Just south of here it cuts parallel to, then through the Milton Moraine in a bedrock valley. In postglacial time, buried ice remained in the Milton Moraine for perhaps more than a thousand years after the main ice sheet retreated. During this time the present valley of the Yahara was blocked by this mass of buried ice that was covered with debris. Because that outlet was blocked, water levels were dammed and water flowed out through the valley that you see to the east of you until enough ice melted to allow drainage in the present location of the Yahara River. Continue east and then south across outwash.
1.2

47.5

Turn left (east) on Leslie Road. All of this flat surface is outwash deposited before the draining of glacial Lake Yahara.
0.9

48.4

It is through here (at the railroad crossing) that drainage from glacial Lake Yahara passed. Note that the outlet has a very low gradient and a very shallow valley, suggesting that water flowed through this valley for only a short time as glacier ice was melting out of the Milton Moraine, which lies to the south.
0.9

49.3

Turn left (north) on Tower Drive. You are entering a ground moraine area with a fairly large number of drumlins in its northern part. The till is thin here, and there are several exposures of dolomite bedrock. Note as you head north that the lineation of the topography is more apparent and that the number of drumlins increases.
1.0

50.3

Continue straight on Highway A. Look to the left at the Yahara lowland.

0.4

50.7

Leave Highway A. Continue straight on Tower Drive.

1.1

51.8

Cross Highway 51.

0.4

52.2

Relatively good drumlins are present to the east and west of the road in this area and to the north along your route.

2.6

54.8

Turn right (east) on Highway B. Look west across the Yahara Lowland to the Milton Moraine. The hill to your right (beneath the radio tower) is a good drumlin. You continue on ground moraine.

1.4

56.2

Turn left (north) on Highway W in Utica. Continue on ground moraine.

1.1

57.3

Just after passing beneath I-90, you enter an area of ice-contact stratified drift composed of gravel that was probably deposited in or under the ice. Note the kettles and kames in this area. Small sand and gravel operations are present in this variable material.

0.8

58.1

Turn right on Drotning Road. Kame and kettle topography is especially noticeable in the next mile. Note also that road cuts show the presence of sand and gravel. Note again how the nature of land use in this ice-contact stratified drift area differs from land use on ground moraine surfaces. This deposit formed in tunnels in and beneath the ice like the deposit you saw north of Brooklyn.

1.0

59.1

Turn left (north) on Evergreen Drive. You continue in the area of ice-contact stratified drift.

0.8

59.9

Intersection with Prairie Queen Road. Continue north on Evergreen Drive.

0.1

60.0

Descend onto a pitted outwash surface. Note the drumlins ahead. You are now entering a landscape that is typical of nearly all of the northeast part of the county. It is a landscape of ground moraine with many drumlins separated by low areas, which at the time of deglaciation carried meltwater away from the retreating ice. Some of the valleys are floored with outwash and some are floored with till similar to that in the surrounding ground moraine. On top of these materials there are often several feet of fine-grained silt and clay deposited in temporary lakes or simply washed from adjacent slopes. Most of these areas were extensive wetlands before they were drained and you will notice many straight drainage channels indicative of drainage for agriculture. Many of the hills in this area are cored with bedrock and we will pass several quarries, most of which contain dolomite.

0.7

60.7

Cross two drumlins grown together and return to the pitted outwash surface.

0.6

61.3

Here you skirt the south end of another drumlin before entering the pitted outwash surface again.

0.6

61.9

Enter ground moraine and turn right (north) on Highway W. You will be crossing ground moraine for the next 6 to 8 miles with numerous drumlins in sight on both sides of the road. The individual drumlins will not be identified.

1.3

63.2

Intersection with Highways 12 and 18. Continue north on Oak Park Road. You have been, and will be, going by numerous drumlins.

1.4

64.6

This area with wooded drumlins is typical of the landscape in the drumlin area of eastern Dane County. All of the area is underlain by till.

1.6

66.2

Turn left on Highway BB. You are still on ground moraine.

0.7

66.9

Here you enter a small recessional moraine deposited as the ice retreated. A set of small recessional moraines in this area was grouped together and called the Lake Mills morainic system by Alden

1.1

68.0

Stop here for a moment and look at the topography around you. To the right and left of the road are low areas that are artificially drained by straight drainage ditches. These areas are covered by organic soils. Several drumlins standing above the organic soils can be seen to the north of the road. The slight ridge you are on is the end moraine and the edge of the ice was parallel to the highway. Ahead of you is a series of drumlins that were formed earlier while the ice margin was at or near the Johnstown Moraine. As the ice retreated, this end moraine was draped over and around these drumlins.

0.6

68.6

You are crossing a drumlin here and there is another one a few hundred yards ahead.

0.8

69.4

Note that here the moraine ridge nearly disappears and that you are on flat outwash surface covered with fine-grained silt and clay and organic soil.

0.6

70.0

Continue onto ground moraine on Highway BB.

1.4

71.4

Turn right (north) on Highway N.

1.2

72.6

You are now leaving ground moraine and entering another small segment of the end moraine that we were on a few miles back. Note the quarries in the vicinity that are crushing dolomite and just ahead there is a gravel pit on the right. This gravel pit is in ice-contact stratified drift.

0.3

72.9

Continue north under I-94 and across Highway TT. Across the valley to the right is a small area of kames and kettles and the gravel pit, worked to produce aggregate for asphalt, is in the sand and gravel.

1.0

73.9

Continue on Highway N across more poorly drained areas with fairly thick silt, clay, and peat, and more drumlins.

1.4

75.3

Turn left (southwest) on Highway T. You are still on ground moraine on high areas and outwash in low areas.

0.9

76.2

Turn right (west) on Burke Road. We continue on ground moraine for several miles.

0.9

77.1

Turn right (north) on Bailey Road. Still on ground moraine.

0.5

77.6

Turn left (west) on Nelson Road. Old gravel pits on right are in ice-stratified drift.

1.1

78.7

As you descend the hill, note the fairly large valley ahead of us. This valley is part of the early bedrock valley system of the Yahara River drainage (see Figure 2 in text). The valley also carried water beneath the ice as it was retreating and may have carried a considerable amount of water into glacial Lake Yahara after most of the ice was gone from this location. Note that on both sides of the road extensive sand and gravel mining has been done. The pits are in kames and kettles that contain sand and gravel. This was deposited on and against ice that was in the valley and that collapsed as ice melted out. Thus, you can see here evidence that glacier ice remains somewhat longer in the valleys than in the adjacent uplands because if the valleys had been icefree, gravel would fill the entire valley and not be concentrated on the valley walls.

0.5

79.2

Turn right on Reiner Road. You are now on outwash in the valley bottom. Drumlins lie to the right and gravel pits in the outwash are on the left. Continue north on Reiner Road, past numerous old gravel pits in outwash.

1.3

80.5

Turn right (northeast) on Highway 151. You are surrounded by fairly featureless ground moraine.

0.3

80.8

Turn left (west) on Highway C and Hoepker Road. Continue straight (west) on Hoepker Road. The topography you are crossing is gently rolling ground moraine, mostly determined by the surface of the underlying bedrock.

2.4

83.2

Cross a bridge over I-90.

0.7

83.9

Turn left (south) on Highway 51. The quarry over the ridge to the west of the road is in Galena-Platteville dolomite.

1.0

84.9

Turn right on Messerschmidt Road. To the south is Truax Field and the deposits of glacial Lake Yahara.

0.7

85.6

The hill you are on is underlain by St. Peter sandstone. To the south, the flat surface of glacial Lake Yahara underlies the runways. The sediments continue to the north under Cherokee Marsh.

0.2

85.8

Turn left on Highway CV. To the north lies Cherokee Marsh, also underlain by deposits of glacial Lake Yahara. In this area organic materials (now drained) overlie silt in the central parts of the lake basin and sand is present relatively close to the surface around the edges. In many cases these sands were deposited as outwash, although in places they have been reworked by waves of glacial Lake Yahara.

0.8

86.6

Continuing south (left) on Highway CV (Packers Avenue), you rise onto ground moraine and can view the flat surface of glacial Lake Yahara deposits to the left.

1.7

88.3

Continue south on Highway 113 (Packers Avenue).

1.0

89.3

Remain on Highway 113 at the intersection of Highway 113 and Aberg Avenue. You leave the ground moraine and adjacent outwash deposits and reenter the area covered by deposits of glacial Lake Yahara. Within the city of Madison most of the deposits have been drained and then filled so that very few remnants of these deposits can be seen on the surface. All of the flat area between here and Capitol Square is underlain by these lake deposits.

1.4

90.7

Intersection with Johnson Street, continue west on Johnson Street.

0.3

91.0

Here you cross the Yahara River which has been straightened as Madison developed. Originally the Yahara River was a very slow moving, meandering stream that flowed across the marsh deposits between Lakes Mendota and Monona.

Small drumlins can be seen ahead just south of Tenney Park and further south near Williamson Street.

0.3

91.3

At the stop lights turn left (southeast) on Baldwin Street and cross one of these drumlins.

0.2

91.5

Turn right (southwest) on East Washington Ave. You will continue on lake sediment until climbing the hill on which the State Capitol has been built. That hill, and the one to the north of it, are both drumlins.

0.9

92.4

Turn left (south) on Blair Street (Highway 151) just as you begin to rise onto the drumlin. Continue on lake sediment.

0.2

92.6

Angle southwest on Highway 151.

0.6

93.2

West parking lot of Law Park and end of trip.1

REFERENCES

1. Willman, H. B. and Frye, J. C., 1970. Pleistocene Stratigraphy of Illinois. Ill. Geol. Surv. Bull. 94, 204 pp.

2. Bleuer, N. K., 1971. Glacial stratigraphy of south-central Wisconsin. Ph.D. Dissertation, University of Wisconsin-Madison, 173 pp.

3. Martin, L., 1965 (1932). The physical geography of Wisconsin. Wisconsin Geological and Natural History Survey, Bulletin 36, 609 pp.

4. Black, R. F., 1960. "Driftless Area" of Wisconsin was glaciated. Geological Society of America Bulletin, v. 71, pt. 2, p. 1827 (abstract).

5. Mickelson, D. M., Knox, J. C., and Clayton, Lee, 1982. Glaciation of the Driftless Area: An evaluation of the evidence. In Knox, J. C., Clayton, Lee, and Mickelson, D. M. (Ed.) Quaternary History of the Driftless Area, Wisconsin Geological and Natural History Survey, Guidebook 5, p. 155-169.

6. Ostrom, M. E., 1970. Sedimentation cycles in the lower Paleozoic rocks of western Wisconsin. In Field Trip Guide Book for Cambrian-Ordovician Geology of Western Wisconsin. Wisconsin Geological and Natural History Survey Information Circular 11, p. 10-34.

7. Dalziel, I. W. D., and Dott, R.H., Jr., 1970. Geology of the Baraboo district. Wisconsin Geological and Natural History Survey Information Circular 14, 163 pp.

8. Cline, D. R., 1965. Geology and ground-water resources of Dane County, Wisconsin. U.S. Geological Survey Water Supply Paper 1779-U, 64 pp.

9. Black, R. F. and Hamilton, T. D., 1972. Mass-movement studies near Madison, Wisconsin; In Morisawa, M., ed., Quantitative Geomorphology: Some aspects and applications. Publications in Geomorphology, State Univ. of N.Y. at Binghamton, p. 121-179.

10. Black, R.F., 1969. Slopes in southeastern Wisconsin, U.S.A. - Periglacial or temperate? Biuletyn Peryglacjalny no. 18, p. 69-81.

11. Black, R. F., 1965. Ice Wedge Casts of Wisconsin. Wisconsin Academy of Science, Arts, and Letters Transactions, v. 54, p. 187-222.

12. Dury, G. H., 1964. Principles of underfit streams. U.S. Geological Survey Professional Paper 452-A, 67 pp.

13. Knox, J. C., 1972. Valley alluviation in southwestern Wisconsin. Association of American Geographers Annals, v. 62, p. 401-410.

14. Davis, A. M., 1974. Pollen analysis and climatic reconstruction: A driftless Area example. In Knox, J. C. and D. M. Mickelson, eds., Late Quaternary Environments of Wisconsin. American Quaternary Association, 3rd Biennial Meeting, Madison, p. 79-85.

15. Dallman, J. E., 1968. Mastodons in Dane County. Wisconsin Academy of Science, Arts, and Letters Review, v. 15, p. 9-13.

16. Dallman, J. E., 1969. Giant beaver from a post-Woodfordian lake near Madison, Wisconsin. Journal of Mammalogy, v. 50, p. 826-830.

17. Maher, L. J., Jr., 1982. The palynology of Devils Lake, Sauk County, Wisconsin. In Knox, J. C., Clayton, Lee, and Mickelson, D. M. (Ed.), Quaternary History of the Driftless Area. Wisconsin Geological and Natural History Survey, Guidebook 5, p. 119-135.

18. Knox, J. C. and Johnson, W. C. 1974. Late Quaternary valley alluviation in the Driftless Area of southwestern Wisconsin. In Knox, J. C. and D. M. Mickelson (Ed.), Late Quaternary Environments of Wisconsin, American Quaternary Association, University of Wisconsin, p. 134-162.

19. Knox, J. C., 1977, Human impacts on Wisconsin stream channels. Annals of the Association of American Geographers, v. 67, p. 323-342.

20. Bleuer, op. cit.

21. Alden, W. C., 1918. Quaternary Geology of southeastern Wisconsin. Geological Survey Professional Paper 106, 356 pp.

22. Black, R. F., 1974. Geology of the Ice Age National Scientific Reserve of Wisconsin. National Park Service, Scientific Monograph Series 2, 234 pp.

23. Alden, op. cit.

24. Florin, M. B., and Wright, H. E., Jr., 1969. Diatom evidence for the persistence of stagnant ice in Minnesota. Geological Society America Bulletin, v. 80, p. 695-704.

25. Alden, op. cit.

26. Upham, W., 1894. The Madison type of drumlins. American Geologist, v. 14, p. 68-83.

27. Whittecar, G. R. and Mickelson, D. M., 1977. Sequence of till deposition and erosion in drumlins. Boreas, v. 6, p. 213-217.

28. Hesler, R., 1978. Sand and Gravel Resources in Dane County Wisconsin. M.S. Thesis, University Wisconsin-Madison, 112 pp.

29. Black, 1974, op. cit.

30. Davis, op. cit.

31. Alden, op. cit.

GLOSSARY

The terms below are defined for the purposes of this report and for the use of the map, Glacial Geology of Dane County, Wisconsin. In some cases, other definitions would be more appropriate if one were talking about the term in general or using the term in other parts of the country.

AGGREGATE - Material consisting of sand and coarser particles that is used as the coarse material in concrete or in blacktop. In either case, the sand and gravel is mixed in various proportions with cement or asphalt.

ALLUVIUM - Sediment that is deposited by a stream. In some cases, alluvium is relatively coarse (if it is deposited by fast moving water), but in most cases in Dane County, the alluvium is made up primarily of sand, silt, and clay.

CHERT - The silicious (SiO_2) rock often left as a residual material after the weathering of dolomite or limestone. Chert is very resistant to weathering and therefore remains a long time in stream systems or on the ground surface.

CLAY - Very fine-grained sediment that in this report refers to the particles less than .002 mm in diameter.

COLLUVIUM - Material that has moved down slope by sliding and surface wash. It is poorly sorted and contains a mixture of the materials formerly present on the hillslope above.

DOLOMITE - A common rock type of Dane County. It is made up of magnesium and calcium carbonate and was deposited under marine conditions several hundred million years ago.

DRIFT - Sediment that is deposited either directly or indirectly because of glaciation. It includes materials deposited directly by the glacier ice (till) and materials deposited by streams flowing away from the glacier ice.

DRUMLIN - An elongate hill streamlined in the direction of ice flow. In Dane County, drumlins are oriented northeast-southwest and contain either a bedrock core, a sand and gravel core, or are entirely made up of till. They form beneath thick ice because of differential erosion or deposition by the ice itself.

END MORAINE - A moraine formed at the terminus, or margin, of a glacier.

EROSION - The process of removal of weathering products by some agent such as glacier ice, water, or wind. In the case of water erosion, materials are moved to a lower position on the slope, into stream systems, and eventually out of the area.

ESKER - A long, sinuous ridge that is formed beneath glacier ice by flowing water. Water produced by melting on the glacier surface or at its base is concentrated in tunnels beneath the ice and occasionally these streams deposit sand and gravel, filling the tube. After glacial ice has melted out, the stream deposit is left as a ridge in the shape of the former ice tunnel.

FLOODPLAIN - The fairly flat area adjacent to streams that is covered by flood water at various times and that is underlain by sediments deposited by the flooding stream.

GRAVEL - Material coarser than sand (2 mm) size.

GROUND MORAINE - A landscape produced primarily by deposition of till beneath glacier ice. The ground moraine surface may include drumlins and in places may be rolling or in other places may be relatively flat. Areas of end moraine are excluded.

HUMMOCKY - Refers to a surface with many closely spaced small hills and depressions.

HOLOCENE - The time period of approximately the last 10,000 years. This is the time of climatic warming after the preceding cold epoch.

ICE-CONTACT STRATIFIED DRIFT - Water-deposited material that is deposited in contact with glacier ice. Because of the relatively short distance of water transport, materials are poorly sorted. Because the deposits collapse as the adjacent or underlying ice melts, stratification is generally disturbed or destroyed. Both factors contribute to the deposits being variable in composition.

KAME - A hill, somewhat equidimensional in shape, composed of ice-contact stratified drift. Kames form when debris washes off the ice and out of the ice into depressions in the ice.

KETTLE - A closed depression, often found in ice-contact stratified drift and in end moraines, that forms when an ice block which is partially or completely buried melts out.

LIMESTONE - A type of sedimentary rock composed primarily of calcium carbonate.

LOESS - Windblown silt that is picked up from a relatively barren land surface as ice retreats and is carried by the wind to its point of deposition. In Dane County, loess blankets all of the glacial deposits in varying thicknesses.

MEANDER - One of a series of somewhat regular and looplike bends in the course of a stream. Stream erosion takes place primarily on the outside of meanders and deposition takes place on the inside.

MORAINE - Usually refers to a ridge of hummocky topography that is produced at the position of a former ice margin. See also end moraine, recessional moraine and ground moraine.

OUTWASH - Stratified sand or sand and gravel that is deposited by streams flowing away from a glacier front. These materials are usually well sorted and typically get finer grained in the downstream direction.

PITTED OUTWASH - Outwash that is partially deposited on melting ice blocks. After the ice blocks melt, pits or kettles occur on the otherwise flat outwash surface.

QUARTZ - A mineral composed of SiO_2 that is very resistant to weathering.

QUATERNARY - The most recent period of geologic time extending from approximately 2 or 3 million years ago to the present.

RECESSIONAL MORAINE - An end moraine produced at the glacier margin as it is receding. Contrast this with terminal moraine.

SAND - Mineral grains with diameters between .0625 and 2 mm.

SANDSTONE - Sedimentary rock composed of sand grains cemented together.

SHALE - A sedimentary rock composed primarily of clay particles.

SILT - Mineral grains with a diameter between .002 and .0625 mm.